The Formula

ALSO BY JOSHUA ROBINSON AND JONATHAN CLEGG

The Club
Messi vs. Ronaldo

The Formula

How Rogues, Geniuses, and Speed Freaks Reengineered F1 into the World's Fastest-Growing Sport

Joshua Robinson and Jonathan Clegg

MARINER BOOKS
New York   Boston

HarperCollins books may be purchased for educational, business, or sales promotional use. For information, please email the Special Markets Department at SPsales@harpercollins.com.

FIRST EDITION

Designed by Angie Boutin
Contents page race car © alisinan/stock.adobe.com and graphic
© mlnuwan/stock.adobe.com

Library of Congress Cataloging-in-Publication Data

Names: Robinson, Joshua, 1986- author. | Clegg, Jonathan, 1980- author.
Title: The formula : how rogues, geniuses, and speed freaks reengineered F1 into the world's fastest-growing sport / Joshua Robinson and Jonathan Clegg.
Description: First edition. | New York : Mariner Books, 2024. | Includes index.
Identifiers: LCCN 2023049606 (print) | LCCN 2023049607 (ebook) | ISBN 9780063318625 (hardcover) | ISBN 9780063318663 (paperback) | ISBN 9780063318632 (ebook)
Subjects: LCSH: Grand Prix racing—History. | Formula One automobiles—History.
Classification: LCC GV1029 .R55145 2024 (print) | LCC GV1029 (ebook) | DDC 796.72—dc23/eng/20231107
LC record available at https://lccn.loc.gov/2023049606
LC ebook record available at https://lccn.loc.gov/2023049607

ISBN 978-0-06-331862-5

24 25 26 27 28 LBC 12 11 10 9 8

To our siblings and No. 2 drivers,
Dan Clegg and Céline Robinson

CONTENTS

The Formula

1

Bahrain, 2022

IN EARLY MARCH 2022, THE private jets and cargo planes that ferry Formula 1 cars around the globe—along with drivers, mechanics, chefs, and personal stylists—converged on the desert kingdom of Bahrain.

It was not the kind of place the stars of the world's most popular motor racing series would normally visit of their own accord. The country is smaller than Luxembourg, hotter than melting rubber, and has a human rights record generously described by activists as "dismal." But Bahrain also pays tens of millions of dollars every year for the privilege of being the first stop on the F1 calendar. As teams geared up for the new season, they were flying in for three days of painstaking tests at the Bahrain International Circuit. This would be their last chance to perfect their 2022 designs before the opening Grand Prix of the year.

The drivers, a collection of slightly built millionaires without normal human fear receptors, filtered in from their respective corners of the world—but mostly from Monaco, their favorite tax haven, on the French Riviera. To them, preseason testing in F1 is a lot like the first day of school. Only instead of everyone leering over each other's backpacks and sneakers, they're checking out bits like wing flaps and diffusers. This year, they'd come to Bahrain all itching for a look at the same thing: the newest design sitting in the Mercedes garage.

What had the richest team in F1 cooked up now?

The intrigue on this March morning was even higher than usual. For the better part of a decade, Mercedes had been the sport's ruling empire, an untouchable paragon of German engineering. Whatever you could do, Mercedes did better, smarter, and in clean silver paint. Not only were they the wealthiest outfit in the sport, they also made everything look so effortless. The team went about its business of obliterating the competition with old-world class and a healthy dose of snobbery. At times, it felt like Merc had more in common with a fancy watch company or a maker of fine leather goods than with the noise of wheel guns and the smell of exhaust fumes. That's what happens when a team laps the field and wins seven world championships in a row. Or at least, that's what *did* happen until the final race of the 2021 season.

Four months before they came to Bahrain, Mercedes had been knocked off the top spot in nearby Abu Dhabi, as Red Bull edged it to the title in highly controversial circumstances. There had been plenty of bluster, and even talk of a legal appeal from the team, but Mercedes eventually dropped the protests and decided to focus on doing what they always did: building a faster car. So by 2022, all of Formula 1 was bracing for the empire to strike back. Mercedes had the smarts and the financing to do it with a vengeance. And most of all, they had people who knew just how to pull it off.

The team's recent dominance had been built around two men. One was its team principal, a suave Austrian named Toto Wolff, who speaks five languages (conservatively) and wears nothing but crisp white shirts (handsomely). The other was its British star driver, Lewis Hamilton, who had won seven world championships and whose wardrobe consists of everything except crisp white shirts.

On this particular morning, Hamilton strode through the desert heat to the Mercedes garage in head-to-toe Alexander McQueen, from his oversized knit sweater to his leather combat boots. In a sport where most drivers are exclusively dressed in team-issue gear and ill-fitting shorts around the circuit, Hamilton treats the F1 paddock like his own personal catwalk. That's because more than a decade at the top has put him in an entirely different atmosphere of celebrity from every single one of his fellow drivers. The others might sport slick watches and fancy sunglasses once in a while, but their primary fashion statement is really just a collection of sponsor logos. Only Lewis has every creative director in Milan on speed dial.

Wolff, who personally signed Hamilton for Mercedes in 2013, hardly noticed the fashion statements anymore. All he cared about was what Hamilton could do behind the wheel. Their partnership had produced five drivers' titles and eighty-two Grand Prix victories, and had made them both unfathomably rich. Whatever worked for Lewis also worked for Toto. "Seven-time world champion," Wolff says. "You hold the records for the most wins and the pole positions, you can show up with a pink suede tracksuit."

Besides, Wolff knew that Hamilton's wardrobe would not be the only unconventional design choice to emerge from the Mercedes garage that morning. And the other one cost a lot more than the latest in haute couture.

For months, Wolff and his team of engineers had been working on a revolutionary new car for the 2022 season, one that pushed up against the limits of the sport's rulebook and marked a departure from forty years of conventional thinking about F1 design. The Mercedes W13 model was such a top-to-bottom overhaul of the previous season's W12 that the only component the team carried over was the steering wheel.

Now, ten days before the opening race of the season, they were about to unveil their creation and test it in race conditions for the first time.

Even Wolff had to admit that Mercedes was cutting things a bit close. Road-testing any new car less than two weeks before the start of the season counted as a major risk. Road-testing a new car that had been developed in secret and represented a radical reinterpretation of the sport's regulations counted as borderline reckless. The final preseason test is supposed to be for fine-tuning and minor upgrades, not drastic revamps and innovations from deep left field.

Wolff understood all of this. But even as he stood outside the Mercedes garage with the moment of reckoning just minutes away, he felt confident in his calculated gamble. Things had a way of working out for him. This was a man who once crashed a Porsche 911 at a track in Germany going 179 miles per hour and emerged without lifelong damage. This was also a man who earned his job at Mercedes by delivering a scathing presentation in which he ripped the company's directors to shreds. In short, Toto trusted Toto.

He liked to think of the F1 season as a game of chess. Pre-season testing represented your opening gambit. You spent all winter marshaling your resources, analyzing your opponents' weaknesses, and coming up with a plan of attack. But once you arrived in Bahrain and the covers came off, the basic outline of the season was set. You could make adjustments and upgrades to the car in the weeks and months that followed, but there were no do-overs.

Which is why Wolff had kept the W13 under wraps until the last possible moment. If Mercedes got this one right, the team wouldn't merely get a jump on the rest of the field, it would leave the others with no chance to catch up. By the time anyone else developed a similar machine, the season could be halfway over, at which point Wolff would have Mercedes back where he knew it belonged: way out in front, with the rest of Formula 1 trailing in a cloud of dust.

Shortly before 10 a.m., Wolff made his move. The Mercedes garage doors went up and out rolled a silver car whose basic geometry sent a roar up and down the pit lane like a V12 engine at full throttle.

The new Mercedes didn't look how a Formula 1 car was supposed to look. Sure, it had four wheels and a giant wing on the back. But behind the driver, where the bodywork flares out to give an F1 car its distinctive teardrop shape, Mercedes had done something wild. The areas where the car curves out are known as the sidepods—they house the cooling system and take in air to keep the engine from overheating. Or at least, they used to. Seeking a dramatic increase in aerodynamic performance, Mercedes had removed the sidepods from its new car entirely.

This was the equivalent of Roger Federer turning up to Wimbledon with a square racket. The fundamental shape of modern F1 cars had been more or less unchanged since the late 1970s. But where the smoothly contoured sidepods normally belonged, the W13 featured a section of heavily sculpted and oddly angled bodywork that was vaguely reminiscent of a stealth fighter.

The new-look Mercedes did not evade detection for long. Seconds after it was wheeled out, a scrum of photographers surrounded the car, while pit crews up and down the paddock looked on in openmouthed shock.

Even the Mercedes drivers had been taken aback by the new look. When George Russell, the boyish, clean-cut Brit who'd just joined the team, first laid eyes on the car he was supposed to drive, he was hit by one lasting impression.

"*Fuck*," he thought. "This looks fast."

Lewis Hamilton, his teammate, had never seen anything quite like it either—and he'd seen pretty much everything over his fifteen years in F1. The car looked so sleek, so streamlined. But it also looked so . . . different. And in F1, Hamilton knew, different can go one of two ways. It either wins you a championship, or it turns you into carbon-fiber pack fodder.

It was that first possibility that worried Red Bull's team boss, Christian Horner, a man who spent as much time fretting over what Mercedes might be up to as what happened inside his own garage. As the longest-serving team principal in the sport—and yet somehow one of the youngest—the forty-eight-year-old Horner was the same kind of pathological competitor as Wolff, only in a completely different package. Instead of being a suave European, he was a squinting Englishman who complained so loudly and so often that Wolff had compared him to a yapping terrier. The last thing Horner needed now was another year of dealing with a super-quick Mercedes. Not only would it make the season feel longer than the ten-month slog it already was, but Horner knew that Toto would never shut up about it. So Horner, a master of allusion and insinuation, began wondering aloud to anyone who'd listen whether the new Mercedes might actually be illegal.

He soon discovered that Merc had already taken care of that.

MONTHS BEFORE BAHRAIN, THE MERCEDES DESIGN GROUP was coming to grips with F1's first major rewriting of the sport's technical rules since 2014. Poring over every detail, the engineers thought they spotted an opportunity. The new regulations, which are normally written with all the precision of a medical textbook, had left something just vague enough about the shape of the car.

Mike Elliott, in his first season as the team's technical director, took it to mean that he had more freedom than usual around the sidepods. For a man with a doctorate in aerodynamics from Imperial College and more than twenty years' experience in F1, this was the moment to shine. If Elliott's instinct proved correct, the Mercedes would be marginally more efficient as it sliced through the air. "We're talking about something that if we'd have got it right," he says, "maybe we would have been three- or four-tenths quicker."

In F1, three- or four-tenths of a second per lap counts as an eternity. So just to make sure that all of this would be okay, Mercedes quietly approached the sport's rulemakers, the Fédération Internationale de l'Automobile, also known as the FIA, with its proposed outline. They took one look at Mercedes's spin on the regulations and let out a collective sigh. The FIA officials were adamant that this sort of extreme interpretation was not what they had intended. But they were less adamant on whether it broke any rules. The only thing they could say categorically was that the Mercedes W13 wasn't *not* legal. Mercedes took that as a ringing endorsement.

Just eight weeks after running the design through its initial simulations, Elliott watched as the W13 lurched down the pit lane and disappeared out of view. He turned to face the bank of monitors in front of him. There are thousands of sensors on an F1 car, recording everything from tire pressure to spring compression. But there was only one number that mattered now. It took barely one and a half minutes to complete a lap at the Bahrain International Circuit. In less than a hundred seconds, Elliott would know just how fast his creation could go.

He waited as the time ticked by. Sixty seconds . . . sixty-five . . . seventy . . . Elliott glanced at the end of the home straight, waiting. Eighty seconds . . . In a flash, the W13 roared back into view. Hamilton rounded a right-hand bend and the Mercedes powered past the pit lane in a silver blur.

Elliott's eyes were fixed on the screen. The lap time flashed up instantly: 1:40.60.

He already knew these weren't ideal race conditions. It was too hot and too windy. Most teams still had their cars back in the garage, waiting for the cool of the afternoon. He knew Hamilton wasn't pushing the car to the limit either, that it was carrying a nearly full tank of fuel and the tires weren't at their ideal temperature. Mike Elliott understood all of this. But deep down, he also understood that 1:40.60 was a sign that Mercedes was in deep shit.

Hamilton crossed the line for a second time. 1:40.45. Hardly an improvement.

The Bahrain preseason test was barely a few minutes old—and entirely inconsequential in the grand scheme of the F1 season, since they don't give out points for testing. But as the morning sun beat down, a troubling thought began to crystallize in Mike Elliott's mind. "I think we made a mistake," he said quietly. "We got something slightly wrong."

Moments later, Hamilton's voice crackled over the radio. And that sinking feeling began to develop into something closer to outright panic.

"Something doesn't feel right out here, man," Hamilton told his team on the pit wall. That was putting it mildly. In his many thousands of laps at the pinnacle of motorsport he'd never driven a car that felt less like a finely honed piece of engineering and more like a mechanical bull.

A few garages down, one man knew exactly what the issue was. His name was Adrian Newey and he'd seen this exact problem decades earlier—and he also knew how to fix it.

The only trouble was that Newey worked for Red Bull.

BEFORE WE GET ANY DEEPER INTO THE COLLECTION OF decisions that goes into building a Formula 1 car, or how the sport attracts hundreds of millions of eyeballs for every race, or pays superstars like Lewis Hamilton on a par with the best NBA players . . . it's important to understand that this whole enterprise can be boiled down to one central question:

Which team best understands aerodynamics?

There are a lot of other things that matter when it comes to building a winner. Stuff like engines and tire wear and whether your driver was partying with supermodels on a Monte Carlo yacht the night before. But the department where championships are really won and lost—where teams truly slice off their share of the $2.2 billion prize pie—is aero.

Aerodynamics is the science of how air flows around a solid object—in this case a $12 million race car. Adrian Newey, the sport's most celebrated designer and a guy who's dreamt up championship-winning cars in three different decades, calls it the sport's "biggest single performance differentiator." Which is why he estimates that he has spent a quarter of his life testing ideas in wind tunnels.

A sixty-three-year-old from England's West Midlands whose personal aerodynamics are unencumbered by even a single hair on the top of his head, Newey understood early on that the most significant obstacle in motor racing wasn't other cars. In order to build the quickest thing out there, he first needed to defeat wind resistance.

How he achieved that was down to some forty years of thinking about every nick, groove, and cranny on the surface of a car as if it were the fuselage of a fighter jet. He figured out not only how air met the car as it picked up speed, but also where that air went off each panel and precisely how it was disturbed. The guiding objectives were to make his vehicles as efficient as possible and to generate enough downforce to keep the car pressed to the track. (Think of it as the opposite of an airplane wing. Those are built to help an aircraft generate lift; F1 cars effectively generate anti-lift to stop them from sliding off the road.)

By the 1980s, Newey was firmly among Formula 1's aero pioneers. Their grip on a new field called "computational fluid dynamics" was tightening all the time. They had more data than ever before. And now their machines generated so much speed and downforce that an F1 car going at 150 mph could theoretically drive on the roof of a tunnel without falling down. All the while, Newey's favorite tool remained his HB pencil. No matter how much tech he had at his disposal, everything always began with freehand drawings at his drafting table.

His sketches for the Williams, McLaren, and Red Bull teams had all, at different times, sparked revolutions in the sport. Newey had imagined pretty much everything there was to imagine about F1 design, and he'd tested most of it too. So as he strolled around the Bahrain pit lane eyeing up his opponents' creations, he could anticipate exactly what may or may not be wrong with them. And though the other teams might have wished they could slam their garage doors shut, none of them could really stop him from taking a look. He walked around with a clipboard peering at wings and flaps and sidepods. That clipboard could instantly turn into one of the most precious troves of F1 analysis if anyone ever got their hands on it—or if Newey actually wrote anything down. The real reason he carries it with him, he confides, is to make other people nervous.

By now, what Newey and everyone else on the paddock knew about the 2022 Mercedes was that it was suffering from a phenomenon known as "porpoising"—a bouncing action reminiscent of dolphins leaping through the waves.

This problem had first surfaced in the 1970s and early 1980s. The engineers then had worked out that the car was being buffeted by two competing effects: on one hand, downforce was pushing on it from above, but on the other, a stiff suspension was bouncing it back up. The motion made driving the cars, quite literally, a huge pain in the ass.

"We knew that the ground effect cars of the past had this phenomenon and we talked about it at the design stage as well," Elliott later told a German car magazine. "We didn't expect there to be no problems at all, but none of the simulations gave any indication of how serious the problem would be."

Mercedes had designed the W13 in a simulator with a level of computing power that would impress NASA. The only question was whether the huge potential performance of the car, which the engineers had clearly seen in the simulator, would be accessible in the real world.

What the team hadn't anticipated was that the simulation couldn't precisely replicate the porpoising action. That's because it's impossible to run the scale model of the car that goes in a wind tunnel as close to the ground as needed without tearing up the tunnel's rolling road belts. And in those minute gaps at the very limit of performance, the mathematical equations that model airflow become hugely unstable. There comes a point, in the end, where the process turns into trial and error.

No one had tried and failed more in this area than Newey.

Now, having seen the bouncing Mercedes up close in Bahrain, he knew one thing for certain: Red Bull was on its way to its first victory of the season, before a single race had been run.

THE MERCEDES GARAGE WAS NOW IN THE THROES OF WHAT can only be described as a total freakout.

They had tried everything. Hard tires. Soft tires. Lighter fuel load. Heavier fuel load. Nothing made a difference. Mike Elliott racked his brains, asking himself the same question over and over. "What have we missed?"

The team had expected the upgrade for Bahrain to make the car up to one and a half seconds faster on each lap. But as

Hamilton and Russell wound their way around the sunbaked track again and again, there was no getting away from it. The new design, which one broadcaster likened to a candy bar melting in the sun, was actually slower.

Toto Wolff sat next to Elliott, staring impassively at the bank of monitors logging the lap times. Wolff had always felt at home behind a screen full of numbers. Long before his career in motor racing, he'd been a successful tech investor in the late 1990s. He'd watched the value of start-ups explode, read the way the wind was blowing, and sold off most of his portfolio shortly before the dotcom crash. Wolff made a killing.

He didn't need any help understanding what the data was telling him now. The Mercedes were more than half a second off the Red Bulls. Over the course of a Grand Prix, that would mean finishing half a minute behind.

"At that stage, I knew we were in trouble," Wolff says. "The stopwatch never lies."

The dropoff was so dramatic that Hamilton's biggest rival, Red Bull's freshly crowned world champion, Max Verstappen, suspected that Mercedes was sandbagging. But this was no ploy. Mercedes wasn't pretending to be slower than it really was. It was just really slow. That would become painfully clear as Lewis Hamilton finished a season without a race win for the first time in his entire career.

In the weeks and months ahead, Mercedes would trace its costly mistake back to one single data point in the initial design simulations. Of course, it was too late to do anything about it quickly. Undoing the misguided approach for the W13, which also influenced the 2023 car, would take nearly two seasons and the best part of $700 million. It's no exaggeration to call it one of the most expensive mistakes in the history of professional sports.

That all of it could come down to such a fine detail—one errant grain of sand on a beach—illustrates what distinguishes Formula 1 from every other major sport. This is a competition where the most decisive action of the season can take place not on the track in the middle of a Grand Prix, but in a wind tunnel simulation that no one sees, somewhere in the British countryside, months before the season begins.

By all rights, this should not make for a compelling spectacle. And yet in spite of all that, Formula 1's global growth has become one of the biggest success stories in twenty-first-century sports.

It's a business where the rules change constantly and even people with PhDs in aerospace engineering can't always figure them out. If soccer became the world's most popular sport because it requires no translation, then F1 is the opposite. Making sense of it can seem like it requires graduate-level physics. Most of the people competing in it fail to grasp all of the intricacies all of the time. Lewis Hamilton can tell you how a car feels in his bones when it understeers or oversteers through a high-speed corner, but he can't describe the inner workings of a fuel injection mechanism. Nor does anyone expect him to. The chemical engineers designing rubber compounds at the Pirelli tire factory know next to nothing about gearboxes, and a team's aero specialists know precious little about a car's electronic systems. So many people are required simply to get an F1 car through a race (let alone win one) that team offices during a Grand Prix look like Apollo Mission Control. Each team brings around two hundred staffers to the circuit and has up to a thousand more back at the factory. Nothing could be further from eleven guys kicking a ball into a net.

Yet far from being a liability, this level of complexity has turned into its main appeal, because it captures the constant and unforgiving high-wire act that is required to survive in Formula 1. This is a sport where success is built on perpetual reinvention. The combination of shifting rules, evolving technology, and astronomical rewards means that it's not enough to do things marginally better than your rivals. The only way to win championships is to land a series of technical moon shots—and then do it all over again. Long before Silicon Valley promised to move fast and break things, Formula 1 was driving faster and rebuilding things every year.

Sometimes this reinvention has come with fantastic, billion-dollar success. Sometimes it has almost killed the sport. And sometimes it has simply produced a really lousy Mercedes. Whatever the outcome, the only constant has been that nothing was ever constant.

That's why this book isn't a blow-by-blow retelling of the greatest moments on asphalt or a completist breakdown of champions and their cars. Instead, it tells the story of disruption, reinvention, and how Formula 1 has remade itself again and again over the course of its existence. That spirit of reinvention took root from the earliest days of the sport, long before it was considered safe, profitable, or fit for television, and eventually set up Formula 1 for global success. But along the way, that same spirit also threatened to drive it off a cliff. That was the price to pay in a sport where bending the rules to the point they almost break is the only business model.

So how F1 built an empire in the twentieth century, nearly became obsolete in the early twenty-first, and reinvented itself in an age of emissions standards, electric vehicles, and on-demand streaming is one of the most remarkable comebacks in corporate history—a case study that pro sports from tennis to golf to the Tour de France have scrambled to replicate. This story of how F1 reached its current heights is both more complicated than executives at Netflix would have you believe and much simpler than its roller-coaster history would suggest.

And if F1 proved that an ancient series could innovate and adapt on the fly, it shouldn't come as much of a surprise. Far from anomalous, F1's present success is merely the latest iteration of the sport's powerful, entrenched philosophy that rewards evolution and experimentation above all else. After all, when risk-taking is in your DNA, you are always on the precipice of greatness, right up until you slam into the wall. The entire sport is about understanding the rules, finding the gaps, and driving through them with your foot to the floor.

2

Loopholes

THE THING ABOUT THE WORD "Formula" is that it calls to mind a mathematical equation or a chemical compound, not an open-wheel motor racing series. Yet the term hasn't changed since the championship was founded in 1950, because it refers to the dense set of regulations and specifications that every car and driver must follow in order to race in a Grand Prix.

In other words, Formula 1 is a sport literally named after the rulebook.

Which is odd, because no other sport spends as much time rewriting the entire thing as F1. The laws of soccer were drawn up in the back of a pub during the reign of Queen Victoria and have barely changed since. Golf has been golf ever since some windblown Scotsmen laid out the rules in 1744. And baseball, despite its constant fiddling at the margins, still looks much the way it did in Babe Ruth's rookie year.

Even the leagues that do make major changes usually hold themselves to about one seismic shift per century. When American football introduced the forward pass or the NBA implemented the three-point line, those revisions to the rulebook table-flipped the entire sport. They upended how the games were played in ways expected, unexpected, and absolutely unthinkable—and they reverberated for decades afterward.

Formula 1 does that every couple of years.

The reason is that F1 is as much a technology problem to be solved as it is a competition to be won. It constantly updates itself to stay at the forefront of elite car development and to prevent the competition from becoming static—change and evolution are the entire point of the sport.

Whole seasons of work become obsolete overnight, as teams are sent back to the drawing board by someone else's bright idea. All of which means that quick-thinking engineers and visionary designers are as important as the lunatics behind the wheel. The drivers may be the ones who get doused in champagne and invited to the Met Gala, but the real game-changers are the nerds who spend their entire careers studying the rulebook and hunting for loopholes. Speed on the track comes first and foremost from speed of innovation.

The history of the sport, from its very beginnings, is littered with secret weapons that win just enough races before they're finally outlawed. They have insane-sounding names that fall somewhere between geometry instruments and dishwasher parts: parallelogram skirts, active suspension, and double diffusers. Those are all real, by the way. And each of them led to a championship-winning advantage for their teams before anyone else figured out what was going on. They were also the product of some serious reinterpretation of the rules.

There's just one problem with loophole technology: it never lasts long. Once other teams work out the trick, they are shameless about trying to replicate it. When the legendary South African engineer Gordon Murray saw a rival team come out with a performance-enhancing piece of design, he didn't think twice about how to react. "Some people resent them," he once said. "I never do. I just get on and copy it."

This divides F1 designers into two distinct classes. You're either the one that finds the loophole and makes it look like a cheat code, or you're stuck in the pack frantically trying to catch up.

The only alternative is to appeal to the FIA, the keeper of the sport's rules. But the process of getting the governing body to close a loophole is time-consuming, expensive, and rarely works—

unless, of course, you're Ferrari. The Italians had such a good record of shouting *Basta!* anytime they saw an innovation they didn't have that the joke among British mechanics was that FIA stood for Ferrari International Assistance.

That's why the first thing a designer like Adrian Newey does before starting work on a new car is spend days, if not weeks, nosing around the rulebook. He steeps himself in what is legal, then starts to wonder what is possible. He focuses on places where the regulations are just vague enough. Or, better yet, he discovers areas of the car's geometry where officials haven't written any specs at all. "An idea pops into your head in the shower or on the way to work," Newey writes. "Ah, there might be a loophole in the rules that helps here."

Once he's figured out what that advantage looks like, it has to work on at least two different cars for more than twenty races a season. At every Grand Prix, each team fields two drivers who rack up points based on where they finish. Their individual totals count toward the Drivers' World Championship, which comes with all the glory. Their combined totals, meanwhile, count toward the Constructors' World Championship, a ranking of the teams that comes with all the money. Winning the constructors' title today is worth north of $130 million. So loopholes don't just make you faster. They also make you substantially richer. If you aren't pushing the envelope, you don't stand a chance—and that's as true now as it was in 1950, when twenty-six cars lined up at a circuit in Northamptonshire for the first ever Formula 1 Grand Prix.

"What one's actually looking for the whole time is the unfair advantage," Williams cofounder Patrick Head says. "The unfair advantage will give you a year's lead—not a week or two weeks' lead, but a year's lead."

NO ONE UNDERSTOOD THIS BETTER THAN A MUSTACHIOED former Royal Air Force pilot whose first love was airplanes. His name was Colin Chapman, and over more than twenty years, he did more to influence F1 design than any other person in history.

This was the man who turned Formula 1 from a playground for gentleman mechanics into rocket science.

All of it stemmed from his singular purpose in life, which was to extract every ounce of performance from whichever engine he happened to be dealt. What Chapman realized first—and far more dramatically than any of his contemporaries—was that there was more to this racing business than pure horsepower. Engine capacity didn't matter if he couldn't first solve the paradox of making his cars as light as possible while also engineering them to stick to the circuit.

His mantra was simple. "Adding power makes you faster on the straights," he said. "Subtracting weight makes you faster everywhere."

Chapman's laboratory was the Lotus racing team, which he founded in 1952 with £25 he'd borrowed from his girlfriend in the North London neighborhood of Hornsey. The area, where his father ran a pub by the railway station, had sustained heavy bombing by the Luftwaffe during World War II and didn't seem a likely place to produce an engineering maverick. The only spot Chapman could find for his first garage was a block of stables full of empty beer casks and crates behind his dad's watering hole. But as it turned out, a unique set of circumstances was brewing all over drab patches of Britain that would turn the country into a cradle for high-level motorsport—and it had a lot to do with six years spent fending off the Germans.

As the dust settled after the war, Britain remained in a state of deep deprivation and under a national rationing program. What it did have in ample numbers was dozens of disused RAF airfields and a supply of war veterans who knew their way around an engine. It was no coincidence that by 1952, when the UK had three permanent racetracks at Goodwood, Brands Hatch, and Silverstone, two had been wartime airfields. It also helped that Britain's newly elected Conservative government had ended petrol rationing two years earlier.

Not only did the country's crop of amateurs now have places to race their cars, they had fuel to put in them too. And in the stables behind the pub behind the train station, Chapman was in his element.

His Lotus road car business quickly hit a wave of popularity, and, even more encouraging, his race cars were actually winning. Chapman grew his engineering ranks by recruiting help from de Havilland, the aircraft company that had produced the iconic Mosquitos for the RAF, and the principles of airplane design were never too far from the stables. What was a race car anyway if not an upside-down airplane wing? Airflow and weight were everything. This being an era before easy access to wind tunnels, the Lotus boys had to come up with creative solutions to measure how their inventions minimized drag. For one test, they fitted fins all over a car to see how they flexed at high speeds. And the only way to get a close look at them in real time, one engineer decided, was to strap himself to the hood for a lap and pray he didn't end up as roadkill.

Chapman's gang would stop at nothing to be just a little bit quicker, and soon everyone knew it. At least one rival referred to them as "that mad lot up in Hornsey," a slight that the intensely proud, easily offended Chapman would not have taken well. A short man with a big chip on his shoulder, he devoted much of his life at Lotus to proving people wrong. The best way to do that in 1958 was to enter a team into Formula 1, even though he doubted the company was mature enough. Chapman simply couldn't bear to see another Englishman named John Cooper—"That bloody blacksmith," he called him—locking up all the best drivers Britain had to offer.

The jump to F1 came with plenty of growing pains. But by the early 1960s, Chapman and his crew had begun to figure out what it took to be competitive. For one, Team Lotus came around to what was known as "mid-engine" design, which really meant moving the engine behind the driver from its regular position in front of him. This might not seem so unusual, until you remember that the natural order of ground locomotion until that point had always been a horse or a locomotive *pulling* the rest of the show, not pushing it.

That was only the beginning. At a time when race car design was breaking free from the confines of traditional geometry, Chapman let his imagination run wild. Ideas poured out of him so fast that as soon as they were shuffled into development, Chapman was already on to the next thing. That manic energy, occasionally boosted by a regime of uppers, could make him impossible to work for. "I know you believe that what you think I said is what I want," read a sign planted on his desk, "but are you sure that what you heard is really what I meant?"

However chaotic the process, Chapman spent the 1960s producing a string of masterpieces, each with its own game-changing innovation. When most cars still used frames where the structure sat on a skeleton of steel tubes, Chapman produced the Lotus 25, which is credited with pioneering the monocoque construction. The French term, which translates to "single hull" and was inspired like everything else by airplane design, meant that the entire chassis, or central frame of the car, was (more or less) one piece of molded aluminum. In Chapman's eternal quest to cut weight from his cars, this was the equivalent of a three-week juice cleanse.

There was, though, one thing Chapman was prepared to *add* to his car before anyone else, and it came from his lifelong struggle to round up money to go racing. In this case, it was a sponsored paint job.

For nearly two decades after the series's conception, F1 cars were simply painted according to nationality. Each country had its own color for handy identification—never mind that most pictures and broadcasts were still in black and white. The French raced in *Bleu de France*, Germans in silver or white, Americans in blue and white, and Brits in what came to be known as "Racing Green." The Italians, in a move that would shape sports car aesthetics, movie posters, and automotive fantasies for years to come, were assigned the color red.

The end to all of that came in 1968, when F1 authorities first allowed sponsored displays on the bodies of the cars. Lotus wasted no time in dumping British Racing Green for the red-and-white livery of Gold Leaf Tobacco. Not only did the deal fill the team's coffers, it also kicked off a forty-year spell of unbroken association between F1 and the tobacco industry. The likes of John Player, Marlboro, Rothmans, and Camel would become as much a part of the sport as slick tires and fireproof suits.

Still, Chapman was nervous enough about Lotus's new look that the team debuted the paint job as far away from prying eyes as possible at a racing series in Tasmania, Australia. Decades before F1 bloggers could instantaneously freak out about the size of sidepods, Chapman figured that unveiling a controversial new paint job ten thousand miles away from Europe would at least buy him some time.

Most important, the new source of income was enough to keep the lights on a little longer in the Lotus garage, where Chapman continued to live up to his reputation as F1's resident mad scientist. Like all mad scientists, he insisted he wasn't mad, simply misunderstood—in particular by the ever-expanding Formula 1 rulebook. In his mind, the whole set of regulations should have been able to fit on the back of an envelope. "Pick the maximum capacity of the engines, choose the type of fuel and specify that the resulting car has to fit in a box so long, so wide, and so high," he said. "Then we can really get down to making a car."

Not that the regulations slowed him down much. Chapman was so relentless in his pursuit of an edge that his own drivers began to wonder if the part of the car he cared least about was the one strapped into the cockpit. Tragedy was such a regular occurrence in Formula 1 at the time that two dozen drivers were killed in the first fifteen years of the series alone. Three had been at the wheel of Chapman's Lotuses.

At the Spanish Grand Prix in 1969, the team nearly added two more to the grim tally when it deployed a version of the high rear wing—an unsightly contraption that sat over the car like a sunshade on a pair of metal stilts, mounted to the suspension. The wing produced tremendous downforce, but it also generated tremendous stress on the arms holding it up.

Lotus ace Graham Hill learned that the hard way on Lap 9, when one of the arms snapped as he hit a bump on the circuit, sending him straight into the metal barriers. Eleven laps later, the same happened to his teammate, Jochen Rindt. The wing broke off, the downforce went with it, and he felt his car practically take flight into the same barrier. As Rindt made his way back to Switzerland with a sore head and covered in bruises, he was incensed.

"I have been racing F1 for five years and I have made one mistake," he wrote in a furious letter to Chapman. "Otherwise I managed to stay out of trouble. This situation changed rapidly since I joined your team . . . Honestly your cars are so quick that we would still be competitive with a few extra pounds used to make the weakest parts stronger, on top of that I think you ought to spend some time checking what your different employees are doing.

"Please give my suggestions some thought," Rindt added. "I can only drive a car in which I have some confidence, and I feel the point of no confidence is quite near."

Sixteen months later, Rindt was dead. A brake shaft in his Lotus 72 failed at the Italian Grand Prix in Monza and sent him hurtling through a poorly installed crash barrier, where the impact caused his own seatbelt to slit his throat. (Rindt still had enough points to win the world championship posthumously.) In keeping with Italian law, authorities charged Chapman with manslaughter as the man responsible for the car. He was acquitted six years later.

"That man," Chapman's British rival Ken Tyrrell said, "should have his own private graveyard."

TRAGEDY DIDN'T DETER CHAPMAN. NOR DID IT TURN OFF MANY drivers from wanting to race one of his speed machines. In fact, by the time a dark-haired Italian American named Mario Andretti got his turn with a full-time F1 contract, he'd spent more than a decade waiting for Colin.

The two had previously worked together for a few Grands Prix in the late 1960s, when Andretti was a young hotshot tearing through the North American racing series and making far more money than he could expect in F1 across the pond. But by the mid-1970s, Andretti had made the leap with Parnelli. The only problem was that the Parnelli team was fast going bust. That's when Andretti and Chapman found themselves sitting in silence at separate breakfast tables in a hotel in Long Beach, California, one day in 1976. And neither was in much of a mood for orange juice. Parnelli had just announced it was pulling out of F1 without bothering to warn Andretti. And Lotus, in a rare slump, hadn't won a single Grand Prix since 1974. As long as the two men were moping, they figured they might as well mope together.

"Mario, I wish I had a decent car for you," Chapman told him.

Mario wished that too. But he also knew who he was dealing with: Chapman wouldn't stay in the dumps for long. Andretti signed with him for the rest of the season and finally broke Lotus's losing streak with victory in the final race of the year at the Japanese Grand Prix. His reward was a contract for 1977 and a personal promise from Chapman.

"Next year's car," he told Andretti, "will make this one look like a London bus."

That extremely not-a-bus prototype would change the course of Formula 1 forever. It was the first of a generation of cars to fully harness the power of something called ground effects.

(If that sounds like sci-fi, you should know that Chapman's most famous involvement with a car was in fact with an even more futuristic enterprise known as DeLorean, which built the gull-wing machine that ferried Doc Brown and Marty McFly through time in the *Back to the Future* movies. Chapman's association with John Z. DeLorean would eventually land him in hot water, but more on allegations of cocaine trafficking and defrauding the British government later.)

For his 1977 car, Chapman was reinvigorated by an eruption of fresh ideas on how to keep a Lotus rubber-side down. More grip meant being able to carry more speed around the bends. And if he got it right, the car would feel like it was cornering on rails. Chapman had laid out all of his thinking in a rambling twenty-seven-page memo to the Lotus R&D department in late 1975.

The crux of it was an insight that no one else had fully considered yet. Bolted-on front and rear wings generating downforce on the car from above were one thing. But, Chapman wondered, what if he could add to that effect from underneath the car?

Lotus designed a system of sliding skirts under a wedge-shaped car that would create a flexible seal with the track. The upshot was that as air flowed over and around the car, the bottom of it would actually be sucked closer to the road. No one had ever seen grip like it.

The concept worked so dramatically that when Andretti asked if he could try it out at a Grand Prix in late 1976, Chapman responded with a firm no, because he didn't want the other teams seeing what fresh hell he was about to unleash on them. Once Andretti finally got behind the wheel, though, he said that the Lotus 78 might as well have been "painted to the road."

That season, Chapman's latest creation was by far the quickest, most advanced car on the grid. By all accounts, including his own, Andretti should have cruised to a world championship. Except the team ran into two issues. One was reliability: "They were built so fragile in every way," Andretti remembers. The other was that Chapman couldn't help himself from being a control freak. He was so maniacal about weight that even the car's fuel load was measured down to the last few drops. The instruction to the race engineers was to set it up so that his machines finished the race with less than one liter left sloshing around the tank.

That was a little close for Andretti's taste. So Mario got in the habit of asking his mechanics for an extra half gallon just before the start. It didn't take long for Chapman to find out about this insubordination. When he did, moments before the South African Grand Prix, he immediately dispatched a mechanic to siphon out the extra fuel. The racing weight was the racing weight, and no driver was going to change that. Five races later, Andretti was leading the Swedish Grand Prix with two laps to go when he ran into a problem with the fuel metering, which meant he was burning through his tank quicker than expected—and thanks to

Chapman, he had no wiggle room. Andretti had to make a pit stop for an extra splash of fuel and finished fifth.

The world championship that the ground-effects car deserved had to wait, then, until the following year. Andretti started the season still driving the Lotus 78, but the even more sophisticated 79 would come online soon. "I was like an expectant father every year, waiting for the child that was going to be the best one yet," he says. "Now, not all those children were better. Some had a short leg or were cross-eyed."

But the Lotus 79 was the closest thing he'd ever driven to perfect. Once the car became available for the Belgian Grand Prix, Andretti won four of the next six races and put himself in control of the title hunt. He clinched the championship that September in Monza.

The celebrations, however, were muted. Andretti's Lotus teammate Ronnie Peterson had been involved in a fiery crash at the start of the race and died that night. It was roughly the fortieth recorded fatality in twenty-eight years of Formula 1 racing—and the fifth Lotus driver. That tragic championship would be Chapman's last. But his worst decisions still lay ahead.

This is the part of the story where John DeLorean makes an appearance.

A smooth-talking former General Motors executive from Detroit, DeLorean approached Chapman with a couple of proposals in the late 1970s. One was to acquire the entire Lotus group outright, which didn't sit quite right with Chapman—though he did need money. The other was a larger scheme to build DeLorean cars in Northern Ireland by using grants from the British government. Specifically, the DMC-12 (flux capacitor not included). With the UK facing mass unemployment, the government signed up to the tune of tens of millions of pounds, completely unaware that DeLorean's company was a financial house of cards. Lotus would help design the chassis for a fee of $17.67 million.

Only none of that money ever found its way to Lotus for developing the car, which turned out to be a leaky, malfunctioning mess. The DMC-12 sold almost nowhere, ran badly, and failed various safety and emissions tests. Worse for DeLorean and Chapman, British prosecutors had been tipped off to their whole arrangement. They later alleged that DeLorean had pocketed $8.5 million of the government fee while Chapman had squirreled away some $8 million for himself.

In a final twist, in the fall of 1982, DeLorean was arrested in a sting operation in Los Angeles while attempting to buy 220 kilograms of cocaine with intent to distribute. (He was later acquitted when a jury found that he'd been the victim of entrapment.) But in England, Chapman could feel the walls closing in. Weeks later, wrought by stress, he died of a heart attack.

Chapman, potentially facing a decade in prison, was fifty-four.

One story that circulated at the time was that he'd actually faked his own death and scampered off somewhere with the money. But there's no evidence that he'd done anything other than keel over. Even with six drivers' titles, seven constructors' championships, and half a dozen revolutionary designs, a staged heart attack and subsequent disappearance was one bit of engineering beyond the talents of Colin Chapman.

His legacy was generations of engineers who made it their life's purpose to bend every rule in Formula 1 in the name of going faster. Their diligence, sharpened by relentless competition, would put the sport on the cutting edge of automotive creativity. Prototypes went from their sleep-deprived brains straight onto the tarmac in a matter of weeks. If they could imagine it and it wasn't expressly forbidden, then they raced it for as long as they could until it was legislated out of existence. That's how the Tyrrell team rolled out a model with six wheels, March developed something called a "surfboard" nose, and Brabham experimented with bolting a giant fan to the back of its car.

To anyone who'd ever raced, the men responsible for pushing the boundaries of Formula 1 were certifiable geniuses. Their Italian rivals, however, had another name for them. To the engineers building the red machines, all of those British mechanics were far from artists to be celebrated.

They were merely *garagistas*. Grease monkeys.

3

The Prancing Horse

THE EPICENTER OF FURY AGAINST the *garagistas* was located in the countryside of Emilia-Romagna, ten miles south of Modena, in a sleepy town called Maranello. That's where a grumpy former racer named Enzo Ferrari did most of his seething.

Enzo had spent most of his adult life producing what he considered to be the most beautiful racing machines in the world—even if that wasn't Ferrari's express purpose when he started the company. In 1929, the legal documents to establish it stated that it was to "buy Alfa Romeo racing cars and take part in the races on the national sports calendar." The man who came to be known as Il Commendatore didn't actually manufacture a single car until after World War II.

But his purpose had always been clear. Enzo wanted to race. From the first season of Formula 1, in 1950, Ferrari made sure that his team would be ever-present in the most prestigious series in the world. It needed to be. And more than seven decades later, Ferrari remains the only team to have participated in every season of the competition. Other titans of the auto industry with proud racing heritage would come and go—Mercedes, Porsche, Ford, and Honda—but the Scuderia, which means stable in Italian, was immovable. After all, the entire operation's mystique depended on it. Ferrari doesn't do TV advertising. Nor does it produce very many cars, with only around ten thousand sold each year. Whatever it does sell trades on a history that was built on the F1 circuit, the image of the red car streaking past stands full of *tifosi* in their racing scarlet. The company used that visceral mix of speed and noise to fuel generations of teenage fantasies, giving each one an iconic road model to pin on their bedroom walls. The Ferrari 250 in the 1950s and '60s, the Dino in the 1970s, the Testarossa in the 1980s . . . The deep rumble of a Ferrari engine, each built by hand and personally signed by the craftsman who made it, was the sound people dreamed of buying the moment they made their first million.

It was an empire built on a feeling. The logo, a black horse once used by an Italian fighter pilot in World War I set against the yellow of the city of Modena, is as recognizable today as the Nike swoosh and McDonald's golden arches. Watch the fans flock to Grands Prix, where estimates suggest at least 30 percent of all attendees are there to support Ferrari, and it would be fair to assume that this team must be the greatest, most successful sporting organization ever to exist—the Yankees, Lakers, and Real Madrid all rolled into one, sweeping up championships season after season.

Only the reality for this company, which spends the bulk of its time selling its own history, is that most of its years in Formula 1 aren't worth remembering at all. That's what was really eating away at Enzo Ferrari in the 1960s as he watched the British *garagistas* stack up world titles in cars somehow faster and more advanced than his. Despite some early success, and not being an engineer himself, Ferrari was so obsessed with his reputation for building engines that he refused to recognize they were only one piece of the puzzle. Colin Chapman at Lotus, who understood that F1 was a sport of permanent evolution, was designing airplane-inspired chassis and exploring the outer limits of downforce. Enzo, meanwhile, still thought that moving the engine behind the driver was some sort of crime against nature. Like a horse pulling a cart, he said, the engine belonged in the front.

Though Ferrari hated so many of the Brits, he had to recognize that this nation of *garagistas* did produce some damn fine drivers. Enzo's preference would always have been to have Italians behind the wheel of his Ferraris, but a Yorkshire boy named Mike Hawthorn brought him an F1 title in 1958, and a former motorcycle champion named John Surtees delivered Enzo another in 1964. What Ferrari couldn't know then was that his deals with the enemy, a rare concession to pragmatism, were really deals with the devil. By letting go of a piece of its Italianness, his national symbol and an export that would become as famous as spaghetti and tiny coffees was giving something away forever. After the 1950s, the Scuderia would never have an Italian champion again.

Not that it prevented the mystique of the Prancing Horse from spreading like a petrol fire. That's because years earlier, in the aftermath of World War II, Ferrari had already nailed his timing more precisely than any cornering driver ever could.

As the fashion for sports cars spread across Europe and the gloom began to lift, Enzo's team rolled onto the scene with bright red paint jobs, elegant lines, and an engine that growled with downright menace. Ferrari and his outfit were crafting handmade machines that won races right as the world was ready to pay attention again. After selling his first pair of road cars in 1947, Enzo knew that only victories would generate more profits, which he could then turn around to finance more racing.

And once Ferrari got started, his team won not only *when* it mattered, it also won *where* it mattered. By taking the checkered flag at a domestic race in the wealthy industrial city of Turin, the Scuderia caught the interest of the upper crust of Italian society just in time. Enzo soon had orders from counts, contessas, and at least one exiled Russian prince. Then in the space of two years, Ferrari won its first Mille Miglia—the prestigious thousand-mile race across Italy—and grabbed first place at the 24 Hours of Le Mans with its speedboat-shaped Ferrari 166. Enzo's great friend, the driver Luigi Chinetti, had been at the wheel for all twenty-four hours after purposely getting his British teammate blind drunk the night before. Even with some light chicanery, the plan to build a car that could stand for speed, style, and performance was working.

By 1951, more than seventy automobiles had rolled out of Maranello and into the princely garages of Europe and Asia. The

Aga Khan, King Leopold of Belgium, and Crown Prince Faisal of Saudi Arabia were all placing orders for Ferraris to call their own. The pace only intensified once Ferrari won back-to-back F1 world championships with Alberto Ascari in 1952 and 1953—its only titles won by an Italian driver. In the sport's era of myth-making, Enzo found himself suddenly in the business of minting legends. Everyone wanted a piece of the Prancing Horse. Ferrari had elbowed its way into some room between Alfa Romeo and Maserati, while Enzo's future rival Ferruccio Lamborghini was still building tractors, not sports cars.

But as cosmopolitan royals gave Ferrari a certain stately patina in Europe, there was one other group of masters of the universe desperate to preach the gospel of Enzo. They came with a thirst for status, a truckload of disposable income, and high, high reverence for the man himself.

This bunch was known as Americans.

From the studio lots of Hollywood to the Ferrari showroom on East 61st Street in Manhattan, anyone with money and a vague appreciation of European luxury now wanted to drive Italian. They spent small fortunes to buy the cars and large ones to keep them running, as Ferrari reps across the Atlantic charged ludicrous prices for parts and maintenance with total impunity. Those bold enough to complain directly to Maranello were told confidently that the problem came from them, not the factory—a Ferrari part would never fail. And still, for all the shabby service, customers kept coming back.

As the Ferrari biographer Brock Yates put it, Enzo had been the first to understand the most important rule of selling old-world taste in the United States: "Treat an American like a hick and you'll own him for life."

IN TRUTH, ENZO FERRARI DIDN'T TREAT HIS OWN COUNTRY-men much better—especially if he happened to be signing their paychecks.

One of his oldest friends, dating back to the Alfa Romeo days, was a man known simply as Pepino. Ferrari hired him as his life-long driver, chiefly to ferry him around his collection of mistresses and favorite dinner haunts. Top of the list was Cavallino, the Ferrari restaurant named for the Prancing Horse and situated just outside the gates of the Maranello factory. But no matter where Enzo dined, Pepino ate at a separate table. And until the end of his life, he was under strict instructions to be idling outside when Enzo finished his last bite.

It was perhaps the least glamorous job in Maranello, but driving a Ferrari to shepherd Enzo around northern Italy was substantially safer than driving a Ferrari for Enzo in races around the world.

When it came to the early, often deadly days of Formula 1, the Scuderia couldn't escape the long list of fatalities. The team's first in F1 was the Belgian Charles de Tornaco, who rolled his car in practice ahead of the Modena Grand Prix in 1953. And the tragedies kept coming at alarming speed. Eugenio Castelotti, the man who was supposed to be the heir to Ascari, also died at Modena after being thrown nearly a hundred yards from his car. That same year, a Spanish driver named Alfonso de Portago was killed in a crash at the Mille Miglia after Enzo had insisted he run the race. The wreck killed one other driver and around ten spectators, including children.

By then, death was a fixture in Enzo's life—and not only on the track. In 1956, his son Dino had passed away at the age of twenty-four after a long illness. It left his father changed forever. Ferrari visited the grave daily and turned his dark office in the back of the factory into a shrine to the young man. Anyone who sat down across from Enzo found themselves staring at a portrait of Dino over his shoulder.

That's where Ferrari found himself grappling with yet another horrific season in 1958, when the British driver Peter Collins and Italy's own Luigi Musso were both killed at Grands Prix in Ferrari 246s mere months apart.

"I have lost the only Italian driver who mattered," Enzo said of Musso.

Those bloody seasons put Ferrari in the crosshairs of two of Italy's most powerful constituencies: the sporting press and the Catholic Church. Enzo may have built a shining symbol of Italy, but none of it was worthwhile if it was all to keep killing Italian boys—never mind the foreigners and spectators. How could Ferrari continue to justify it? By this point, the Vatican viewed auto racing as downright immoral. The Holy See's own daily newspaper, the *Osservatore Romano,* compared Enzo to a Roman god with a taste for flesh. "An industrial Saturn," it called him, "who continues to devour his own sons."

Ferrari himself felt the pain, yet made no plans to slow down. Instead, his solution to prevent Italian boys from dying in his cars was simply to stop hiring Italian drivers. The following season, seven different men would race Ferraris in at least one Grand Prix, and for the first time, not one would be from Italy. There was a Frenchman, two Brits, two Americans, a Belgian, and a German named Wolfgang "Taffy" von Trips. (Von Trips would later be killed while driving a Ferrari at the Italian Grand Prix in 1961.) When Enzo, who once claimed he didn't attend races because he couldn't bear to watch his cars suffer, wrote his autobiography, he called it *My Terrible Joys.*

This whole racing thing had proven to be a destructive habit. But it was also an expensive one. No matter how many road cars he sold, Ferrari almost always found his business short of money. It's why, in the mid-1960s, he welcomed a group of men from Detroit into the dimly lit office at Maranello. They worked for Ford, and they'd crossed the ocean to buy Enzo's company.

Ferrari was interested, or at least he pretended to be. The men wined and dined together, while Enzo insisted that his visitors address him as "Ingegniere" (Engineer), because of his honorary degree from the University of Bologna, even though he hadn't really engineered anything since making horseshoes for donkeys in World War I. The Americans played along and felt like they were making progress. But in 1965, the talks hit a major sticking point.

As always, the question was about racing. Who would have final say over the Scuderia?

"If I wish to enter cars at Indianapolis and you do not wish me to enter cars at Indianapolis," Enzo asked the man from Ford, "do we go or do we not go?"

"You do not go," came the reply.

That settled it. *Ciao.*

Enzo had finally found a bunch of Americans whose money he wouldn't take. He sent them on their way and handed them an autographed copy of *My Terrible Joys* for the road.

How prepared Ferrari really was to sell to Ford remains a question of some debate. One theory is that he was just trying to jack up the price for the suitor he was really interested in, the Italian giant Fiat. And when the empire led by the Agnelli family knocked on his door in 1967, Enzo was ready to listen.

The deal took two years to iron out. In the end, Ferrari accepted far less from Fiat than Ford had offered. But for roughly $11 million, he sold off 40 percent of the company he built and retained 49 percent for himself, plus, of course, final say over racing matters. (The deal stipulated that Enzo's stake would revert to Fiat when he died.) Ferrari also carved out a 1 percent share for the family of his old friend, the designer Battista "Pinin" Farina. And 10 percent went to a mysterious young man named Piero Lardi. What only a handful of people around Maranello knew was that Lardi was actually a Ferrari—Enzo's illegitimate son was being brought gently into the fold.

"Following a meeting of the president of Fiat, Dr. Giovanni Agnelli, with the engineer Enzo Ferrari," the opaque press release read, "it has been decided that the past technical collaboration and support will be transformed within the year to joint participation."

With the agreement, Ferrari exceptionalism became fully wrapped up in one of Italy's largest exporters. The mystique had taken root. The red cars stood for cool, class, and now for an entire country.

The Swinging Sixties might have happened in London, but the Brits were driving Mini Coopers and mopeds. Sex on wheels came straight from Maranello. Steve McQueen was buying Ferraris for his personal collection. So were Prince Rainier of Monaco and his movie star bride, Grace Kelly. The Scuderia, which still refused to advertise, had also hit the big screen in the Hollywood movie *Grand Prix*. Shot with Ferrari's cooperation across 1965 and 1966, it served almost as a propaganda film. Enzo didn't even have a problem with one of the main characters, played by Frenchman Yves Montand, being killed in one of his creations at Monza. His only condition was that no Ferrari could be shown losing to a rival car.

He was seeing quite enough of that in real life.

By the early 1970s, the Hollywood version of Ferrari bore little resemblance to what was happening to Ferrari on the track. Away from the silver screen, it turned out that the Scuderia's Formula 1 car was less useful than a bottle of corked Chianti.

Just as troubling, it was barely Italian at all. The engine remained pure Maranello, with more horsepower than anything coming out of the *garagista* teams. But the chassis used composite materials made in England, the tires were American, and the famous engines were being fed by German fuel injection. Enzo couldn't even bring himself to throw in his lot with a fully Italian driver. When he courted Mario Andretti in 1971, promising him that "everything you see here will be working for you," the criticism from the Vatican still rang in his ears.

"I was Italian but I had an American passport," Andretti says. "So if I got killed, he wasn't going to have the Italian government telling him he was killing Italian talents."

That's how far things had slipped. Yet no one inside the cult of Maranello was willing to say anything. Enzo, whose annual addresses to the staff were still full of oaths of faith and Jupiterian proclamations, demanded total loyalty, which really meant total subjugation. If anyone was going to spark an Italian renaissance, it would have to be an outsider who wasn't under Ferrari's spell.

The one who managed it happened to be a moody Austrian with an overbite.

NIKI LAUDA'S FIRST LAPS IN A FERRARI, ON A TEST TRACK IN 1973, told him everything he needed to know about the creaking Scuderia. When the young Austrian hauled himself out of the car, Lauda took off his helmet and went to share his findings with the Old Man. Ferrari ogled his new driver through the dark sunglasses that were permanently affixed to his face, waiting to hear how great the car was. Lauda was blunt.

"A disaster," he said to Piero Lardi, now legitimate enough to act as translator for his father.

Lardi paused. Lauda couldn't say something like that to Mr. Ferrari.

"Why not?" Lauda asked, as he wrote in his autobiography. "The car understeers and doesn't corner worth a damn. It's undrivable."

Lardi shook his head. That wouldn't fly either, even if the Old Man could tell that Lauda wasn't exactly singing the car's praises. With precisely zero Grand Prix victories that season, it was no secret that the Ferrari was basically an Italian lemon.

"All right then," Lauda said. "Tell him the car isn't quite set up properly. It has a tendency to understeer, and the front axle may have to be looked at."

Lardi could work with that much. He delivered the verdict to Enzo in the politest terms he could muster and braced for the volcano to explode. That's when Enzo surprised him. He turned to Ferrari's top engineer and inquired how long it might take to tweak the car to Lauda's liking. When the engineer told him a week, Ferrari turned back to his new driver with a promise—and a threat. The team would make the changes. But, Enzo added, "If you are not one full second faster this time next week, you're out."

Lauda knew that Ferrari meant it. This was not a man who put on airs, something Lauda understood from that first meeting. The creator of the most elegant machines on the road could also be an absolute boor in private. "He would scratch himself in the most unlikely places," Lauda wrote, "and hawk and spit for minutes on end, with obvious relish, into a gigantic handkerchief which, unfolded, was the size of a flag."

Lauda spent the week obsessing over the car with the engineers, micromanaging every detail of the unwieldy machine. In seven days, they made enough progress to scratch out the full second that Enzo had demanded. And within a year, they found much more.

The analytically minded Lauda discovered that in the chaos of Maranello, there were still opportunities for cold-eyed reason to prevail. For one, Ferrari had something few other teams could dream of at the time: the Fiorano test track, built less than a mile from the factory in 1971. Not only did every corner come with cameras to give drivers and engineers the chance to review performance, the track also had finely tuned timing equipment. Somehow, the outfit that could barely reach F1 speeds in qualifying was turning itself into the sport's first data team. Ferrari's new team manager, a blue-eyed, Columbia-educated aristocrat named Luca Cordero di Montezemolo, also brought a fresh level of discipline to the Scuderia's operations, insisting that they focus only on F1.

Between the billions of Fiat lira now flowing through Maranello, a rejuvenated design team, and Lauda's ferocious determination, Ferrari soon made its way back to where Enzo knew it belonged. In 1975, with the Italian flag painted on the top of the car at Fiat's behest, Lauda won Ferrari's first drivers' title in eleven years. When Montezemolo rang Enzo to tell him they were champions, he heard something for the first and only time on the other end of the line: the Old Man was sobbing.

Lauda, for his part, was far less sentimental about the whole thing. After his five victories and three podiums over the season's thirteen races, he called his armful of trophies "useless" and traded Ferrari's silverware to a garage back in Austria.

In return, Lauda got free car washes.

THE HAPPY, VICTORIOUS PEACE AT FERRARI DIDN'T LAST LONG. Mere months after the title, it all went skidding off the road on the far side of the Nürburgring, a brutal, ancient track in western

Germany that had seen some fifty-one motor racing deaths by the summer of 1976.

That's when Niki Lauda nearly became the fifty-second.

Lauda had never been a fan of the Ring, where a lap around the bumpy fourteen-mile circuit in the forest was more than six times as long as a spin around Monaco. Earlier that season, he'd proposed to an F1 drivers' meeting that they boycott the race entirely over safety concerns. Lauda was voted down.

So on August 1, he reluctantly lined up on the grid of the West German Grand Prix in command of the world championship standings and ready to square off against his new rival from McLaren, the British playboy James Hunt. Except the duel never materialized. On Lap 2, Lauda hit a kink in the road surface coming out of a left-hand turn and lost control of his Ferrari at around 125 miles per hour. Before Lauda could correct, the car jackknifed into the barriers on the far side. A squeal of tires, an explosion of chassis pieces, and then a ball of flames engulfed the vehicle as the fuel tanks surrounding the cockpit ignited. The Ferrari bounced and spun nearly three times before another car crashed into it.

Lauda came to a standstill in the middle of the track while his Ferrari literally melted around him.

Four other drivers immediately hopped out of their cars and ran toward the blaze. An Italian named Arturo Merzario got there first and managed to jimmy Lauda's seatbelt loose and yank him free. Lauda was in such dire straits that Merzario believed he was holding a corpse. The Austrian was unconscious and severely burned. He'd spent what seemed like an eternity inhaling toxic fumes that seared the inside of his lungs. His helmet was charred and fusing to his scalp.

Lauda remembered nothing of this. The next thing he knew he was in the hospital, mutilated but alive. His eyelids were gone, as were his hair and several layers of skin on his face and body. Lauda was both suffering and furious that this had happened. When a priest read him the last rites, it only made him more pissed off.

"My God, Where Is His Face?" was the headline in Germany's *Bild* newspaper. The piece below it speculated that Lauda wouldn't be seen in public for at least six months.

But when Lauda reemerged, it hadn't even been six weeks.

F1 didn't stop when Lauda was in the hospital, nor did he expect it to. All he knew was that Hunt was out there scoring world championship points while he lay in bed. The Austrian Grand Prix came and went, and so did the race in the Netherlands. Hunt had finished first and third in those, which convinced Lauda that he'd recovered plenty and it was time to go racing again. Still in excruciating pain, he hopped into the cockpit of his private plane and flew himself to Italy to put in some laps at Fiorano. As he pulled on his overalls, the Ferrari mechanics there thought they were looking at a ghost.

Peering through surgically reconstructed eyelids, Lauda ripped around the test track and then announced that he was ready to race. The team wasn't entirely sure what to make of it. But forty-two days after the accident, in front of the *tifosi* at the Scuderia's home race in Monza, Lauda was back at the wheel of his Ferrari 312T for a Grand Prix.

He was a picture of courage as he gritted his teeth to pull on his helmet and blood and pus seeped through his fire-resistant balaclava. If anyone needed a reminder that the bravery to be a Formula 1 driver bordered on psychopathic, this was it. Only Lauda was scared out of his mind. Ever since he'd arrived for the Friday practice, he was gripped by terror. His heart was pounding, his stomach was a mess, and he threw up everything he ate. Even then, he finished in fourth place.

The craziest thing about his return was that Lauda still nearly won the world championship. Despite the 12 points Hunt had racked up during Lauda's recovery, the two men traveled to the final race of the season in Japan locked in a dogfight. The only real problem for Lauda then was that it rained.

Days of biblical downpour had prompted him to declare the Fuji Speedway race unsafe. And when it proceeded anyway, he drove only a handful of laps before pulling into the pits and stopping his car altogether. He'd almost died once that season and he wasn't going to tempt fate again. Enzo did not take this decision well. Fully functional Ferraris weren't meant to be parked, especially when it meant simply handing the title to McLaren. The relationship between Enzo and Niki would never recover, though Lauda's contract to drive for the team remained in place for 1977.

That year, Lauda won the world championship almost entirely out of spite.

He hated Enzo, and he hated his Ferrari teammate, Carlos Reutemann. Driving the hell out of the Scuderia's last great car before it was overtaken by the ground-effects generation, Lauda clinched the championship with two races to spare, at which point he felt his job was done. He sent Enzo a telegram from Toronto informing him that he would not participate in the final two Grands Prix of the season.

"I was happy that my departure would be like a slap in the face for Enzo Ferrari," Lauda wrote later. He delighted in turning down one of the largest offers the Scuderia had ever made to anyone. He had won Enzo two titles and now he was telling him where he could shove them. "I don't want to stay," Lauda added. "That's all there is to it."

Ferrari was apoplectic—again. The disrespect of the Austrian was staggering. No one had ever stood up to him like that before. But in Enzo's mind, drivers were as replaceable as a set of tires. Around Maranello, Lauda would not be missed.

Enzo found that a South African named Jody Sheckter was far more pliant. Though he was an unremarkable driver in the grand history of Formula 1, he delivered what proved to be the final championship of Enzo's life in 1979. The 1980s would be worse for Ferrari than they were for hairstyles. Lauda was long gone. Montezemolo returned to Fiat headquarters. And Enzo kept living in the past. While the hated British were pioneering ground effects and later soared ahead in the smaller turbo-engine era, the Scuderia flailed around, a prisoner to its own history and Enzo's resistance to change. Ferrari was still obsessing over horsepower instead of building a car where all the pieces worked together.

The engineers building the different parts barely seemed to know each other's names.

Every other team seemed to be more nimble, to be set up more intelligently, in part because they weren't so beholden to the whims of one old man.

"Honda and McLaren were on opposite sides of the world," said designer Steve Nichols, who joined Ferrari from McLaren. "But Honda and McLaren worked better together than Ferrari chassis and Ferrari engine who were in the next room."

ORDER AND ORGANIZATION WOULD NOT BE AMONG FERRARI'S many legacies in Formula 1. Nor would groundbreaking technology or healthy relationships with drivers. But through his sheer immovability and the might of the Ferrari brand, Enzo turned himself into one of the most powerful people the sport had ever known.

His influence was impossible to ignore, which meant his mood swings were also impossible to ignore. Ferrari understood that the mere suggestion of his stable skipping a race could tank ticket sales for an entire weekend. And when he didn't like a rule change, he carried immense weight to push back against it. That's precisely what he did in 1987 when Formula 1 planned yet another sweeping update to impose 3.5-liter V8 engines, against Ferrari's wishes to build larger V12s. You don't need to be a petrolhead to suspect that Enzo hated the idea. Smaller engines in the world's biggest motorsport made no sense to him. F1 needed to be at the absolute pinnacle. If Ferrari could have put a jet engine under the hood of his cars, he would have.

So Enzo threatened F1 with the nuclear option. If they couldn't agree on engines, then the Scuderia would quit Formula 1, give up on Monaco and Monza, and go race in IndyCar instead. The team went as far as building a prototype for Indy, the Ferrari 637, before the sport's authorities backed down. Even as he approached his ninetieth birthday, the Old Man had some fight left in him.

By 1988, however, the whispers around Maranello were that the Enzo era was coming to an end. It was a sign of how badly his health had deteriorated that when the pope himself, John Paul II, made a pilgrimage to visit the pontiff of motor racing, Ferrari was too unwell to greet him. No longer an amoral killer of Italian boys in the eyes of the Vatican, he had lived long enough to become the man who built cars worthy of a papal blessing.

Enzo died that August. But his lessons on how the sport should be run had made an indelible impression, none more so than on a young British team owner he'd met in the 1970s who would go on to build Formula 1's global empire over more than forty years. His name was Bernie Ecclestone.

"Formula 1 is Ferrari and Ferrari is Formula 1," Ecclestone says. "It's that simple."

4

Supremo

BERNIE ECCLESTONE HAD A LOT of reasons to admire Enzo Ferrari.

When he first entered the sport in 1970, as a sharp used-car dealer who'd turned himself into a driver agent, Ecclestone knew that Il Commendatore had been a force to be reckoned with for nearly twenty years. Not only did Enzo wield tremendous influence over the direction of Formula 1 from his lair in northern Italy, but the old man in the sunglasses learned that there was more to this whole racing game than winning races. Ferrari understood that image mattered. A veneer of class mattered. And more than anything, getting your way mattered.

"The sport is on the table," the Old Man once told him, "and the business is under it."

These were values that Ecclestone had understood instinctively since the school playground, where he sold cookies and buns at a markup, having just bought them from the local bakery with the money from his twin newspaper rounds. Growing up in rural Suffolk and later in Kent outside London, young Bernard knew that he'd have to scrap for every penny. It's not as if his parents would ever be in a position to help. His father was a trawlerman and his mother was an authoritarian homemaker. Their first house didn't even have indoor plumbing.

Ecclestone's gift for salesmanship worked well enough to buy him his first bicycle, which he soon upgraded into a passion for motorbikes. And as Kent built back from the damage inflicted by German bombers, Ecclestone became keenly aware of the growing demand for car and motorcycle parts. It didn't take long for him to start dealing in whole cars and motorcycles, and he honed his skills among the sharks on London's Tottenham Court Road. These included tricks of the trade like running odometers backward and buying and selling large bundles of vehicles, which sometimes included cars that didn't exist at all. In that world of thugs and East End gangsters, the five-foot-two Ecclestone had somehow carved out a spot for himself.

His energy—and occasional skulduggery—established him enough to open a sparkling all-white showroom in Bexleyheath, in the outer reaches of southeast London. It was a long way from the Savile Row tailors where he acquired a taste for buying his shirts, or the Mayfair casino where he whiled away evenings at the card tables. But in Britain's post-austerity days of the 1960s, business was booming. The trawlerman's son now carried wads of bills in his bespoke suits and insisted that his place of business be as immaculate as his appearance. He inspected the floors and adjusted the lighting constantly to show off his growing stock of luxury motors. The only nuisance more annoying than a speck of dirt on the carpet was whichever customer had dragged it in.

The caliber of those grubby customers, however, soon improved along with Bernie's reputation. Celebrity connections from the casino were making the trek out to the showroom and turning it into a smart address for sexy cars. In a moment that perfectly distilled the era, Ecclestone once sold a lime-green Lamborghini Miura to the "It Girl" of Sixties London, a model named Twiggy.

Before long, Ecclestone wasn't just selling cars to chic Londoners, he was loaning those customers the money to buy them as well.

As he raked in cash selling cars and collecting interest, his passion for racing never left him—not that he was ever especially brilliant at it. In his days of tearing around former RAF airfields, Ecclestone scared the living daylights out of himself in at least one serious crash. The death of his close friend the driver Stuart Lewis-Evans from burns sustained at the 1958 Moroccan Grand Prix had convinced Ecclestone to give up racing forever.

But he wasn't ready to quit the game entirely. In addition to being his pal, Ecclestone had also been Lewis-Evans's manager and Bernie couldn't resist a return to the agent business to get his motorsport fix. By the mid-1960s, Ecclestone was hanging out in Formula 1 circles and developing a deep bond with the driver Jochen Rindt. The friendship morphed eventually into a business partnership. This one, too, ended in tragedy. Ecclestone says to this day that he never recovered from Rindt's death at the wheel of a Lotus 72. Strangely, though, Bernie only increased his commitment to the sport after the accident. He and Rindt had harbored dreams of buying a team together. Now Ecclestone decided he should forge ahead alone. So before the 1972 season, he bought the Brabham team for £100,000 (the equivalent of less than $2 million today). Ecclestone had finally joined the Formula 1 club.

The team itself was chaotic and Bernie, owner of the pristine white showroom, was appalled by how messy the workshops were. Everything was splashed with motor oil, and mechanics barely knew where their tools lived. This wouldn't do. But being an owner came with a serious perk: it gave him entry to the periodic meetings of the British constructors at the Post House Hotel near London's Heathrow Airport. That's where they would get together to discuss all of the pressing issues in the sport. As the newcomer, Bernie was in charge of making the tea.

It was around this time that Ecclestone also met the most famous car builder in the world. He and Enzo Ferrari didn't spend much time together, but Ecclestone felt that they understood each other. "He would have made a good used-car dealer," Bernie says.

The two men never had a disagreement, mostly because neither man spoke the other's language. Still, Ecclestone and Ferrari grasped early on what racing fans were coming to see. Whether supporters cheered for the red cars or against them, the Ferraris were part of the furniture. The Scuderia's very presence was enough to set F1 apart from any of the other major motor racing series. Despite the existence of that one prototype, there is no such thing as a Ferrari IndyCar. A Ferrari NASCAR is basically an oxymoron. Ecclestone would always remember this. And as it turned out, no one would get richer off the Ferrari mystique than Bernie, a man who was never once employed in Maranello.

In order to make his fortune, which would swell into the billions, Ecclestone admits that there was never a plan, nor did anyone know exactly how rich he was. That was one of two things he wouldn't talk about. "You'll never get me to discuss last night," he once said, "or money." Bernie wasn't like a driver who knew every bump and curb of a circuit's corners. Nor was he like the engineers who could predict how a complex puzzle of airflow, horsepower, and rubber might behave. Yet he understood how to succeed in F1 as well as anyone.

The recipe, as ever, was to exploit loopholes, uncertainty, and anything that wasn't spelled out in binding legalese. And once you've done that, guard those secrets like a Soviet spy. Share only what you are compelled to share and never let anyone peek inside the garage. It's no coincidence that the motor home that followed him around to Grands Prix from the early 1990s became known as the Kremlin.

"Your problem is, you always want things absolutely clear," Bernie once told a close associate. "And sometimes it's better if things are not clear."

Other histories have been written and filmed about Ecclestone— some authorized, some not, and some in cartoon form. When you live into your nineties as the billionaire architect of a massively popular sport, marry three times, and watch your daughters grow into full-time famous people, the public tends to get curious. But everything you need to know about how Bernard Charles Ecclestone built the foundations of modern Formula 1 and ran his three-ring circus for more than forty years can be boiled down to one fundamental skill. As his longtime right-hand man Michael Payne says, "He was wired to make deals."

Ecclestone used that sixth sense for deal-making to revolution-
ize Formula 1 in three critical areas. The first was taking control of
the teams' relationships with the circuits, an administrative matter
that sounds innocuous enough until you realize that it amounted to
a coup d'état. The second was harnessing television, the single force
on which every modern sports empire is built. And the third was
seizing on the importance of sponsorship—specifically from the
tobacco industry.

As he slowly tightened his grip on all of those, applying the eye
for loopholes and the disruptive streak of a team owner, Ecclestone's
purpose in life became to grow Formula 1 away from the track on a
planetary scale. And if it happened to earn him a pile of cash along
the way, well, did you really expect Bernie to work for free?

His gift was making sure there were always plenty of people
around to foot the bill.

FROM HIS VERY FIRST MEETINGS WITH F1'S OTHER TEAM OWN-
ers, Ecclestone sniffed opportunity. Some of the people around
the table had been hugely successful in the auto industry, and
others were brilliant designers or racers, but if he was being
honest, Enzo aside, these weren't exactly sharks. They weren't as
crafty as the gamblers he met on his regular trips to the casino to
play chemin de fer, nor were they as cutthroat as the gang of used-
car dealers he knew back in London.

Worst of all, none of them had any real money. For all of their
technical acumen, Ecclestone couldn't believe how frequently and
bitterly the teams complained about being on the verge of bankruptcy.

The reasons soon became clear to him. The constructors,
without a lick of commercial sense between them, couldn't agree
on anything. All of it stemmed, Bernie thought, from their failure to realize a fundamental reality of Formula 1: the teams were
only really rivals for a couple of hours a dozen times a year on the
track. They needed to understand that the rest of the time, they
were business partners.

Instead, every outfit from Colin Chapman's Lotus to Enzo Ferrari's Scuderia was out there acting alone for their own small-time interests. Teams negotiated appearance fees individually with each circuit. Ferrari commanded the highest payments, but the *garagistas* were turning up all over the world for just a few hundred pounds. They didn't discuss their finances with each other either. If they thought they'd make up the shortfall in prize money, they were sadly mistaken: the races in those days offered purses of barely $10,000.

No one was quite sure who was running the show either. The F1 calendar was an ever-changing mishmash of races controlled by local automobile clubs, each with their own small-time foibles and sponsor entanglements. The schedule was so chaotic that promoters couldn't even guarantee that every team would show up to every race. The 1969 season, for instance, featured eleven Grands Prix and four nonchampionship races that didn't award points. The following year, there were thirteen Grands Prix, plus three nonchampionship races, and only seven of the twelve teams competed in every single one.

To a stickler like Ecclestone, the whole ad hoc, money-losing setup was nothing short of outrageous. What was the point of ferrying cars, drivers, and mechanics as far away as Canada and South Africa if the team was only going to come home several thousand pounds lighter? What the teams needed was to work together, controlling the terms and negotiating as one. They could be a cartel if they wanted—they just didn't know it yet.

Ecclestone proposed that the British outfits form a company that would optimize the logistics of racing in Formula 1 and make sure everyone got paid fairly, on time and in full. The other owners shouted Bernie down at first. They had no interest in adding more negotiations and more administrative nonsense to their workloads. "Well, I'll do it," Ecclestone suggested. "But I want a fee."

Remember this moment: Ecclestone was offering to take on all of the responsibility and 100 percent of the risk to guarantee proper appearance money for the teams wherever they raced. It's the precise instant when the sands of the entire sport began to shift. Ecclestone would make the deals with the circuits and he would make sure the whole show turned up. All Bernie wanted in return was 8 percent of whatever the promoter shelled out. To the group of teams that would eventually form the Formula One Constructors' Association (FOCA), it was a more than acceptable price for the luxury of never worrying about this stuff again.

At the end of 1972, his first season in the game, Bernie was onto all of the European promoters, demanding they increase the prize fund to £43,000. And in 1974, he even tried his hand at promoting an entire Grand Prix in Belgium after the local organizers proved too incompetent. By 1976, the demands went up to £150,000 a race, £165,000 in 1977, and £190,000 in 1978. If circuits weren't prepared to pay, then the teams just wouldn't go.

Much to the irritation of every official in motorsport, this threat was about to become a recurring theme. But Ecclestone had no problem making himself an annoyance, because he knew everything always came down to money, even when he acted like it didn't. Enzo had taught him that much.

"You should never let people know you're running a brothel," Ferrari once said to Ecclestone. "You have to pretend it's a hotel and keep the brothel in the basement."

Bernie's role at FOCA would see him spend the decade wrangling the strange, confusing world of motorsport politics. His great ally in this was an upper-crust British lawyer named Max Mosley, whose refined upbringing and polished education were everything Bernie's background was not. With a family tree studded with barons and dames, Mosley had been raised in France, Germany, and at British boarding school, all paving the way to Oxford.

"Max could speak English properly," Ecclestone says. "I was a used-car dealer, he was a barrister."

The part of his breeding that Mosley had less to boast about was who his parents were. Max's father was Oswald Mosley, the leader of the British Union of Fascists in the 1930s, whose blackshirted supporters hoped to emulate Nazi Germany and Mussolini's

Italy. Oswald had married Max's mother, Diana Mitford, in a private ceremony at the home of Joseph Goebbels, Hitler's chief propagandist. (This connection would prove especially unfortunate in 2008, when the *News of the World* tabloid obtained footage of Mosley engaged in what it called a "Sick Nazi Orgy with Five Hookers." Mosley successfully sued the paper for a breach of privacy, admitting to the extracurricular activities but denying that there were any Nazi themes.)

Long before Mosley was throwing exotic parties, however, he proved to be the perfect foil to Ecclestone. He had retired from driving in 1969 and cofounded the March F1 team that year, which qualified him as a member of FOCA. Traveling the world and sitting in meetings together, the pair became fast friends.

While Bernie drove the hard bargains by sniffing out any sign of weakness, the less impulsive Mosley considered the long game. Still, he couldn't help but admire his new partner in F1. As an old lawyer of Ecclestone's once told Mosley, Bernie had "a great talent for getting himself out of trouble—that he got himself into in the first place."

Mosley saw it firsthand as he joined Ecclestone to represent FOCA's interests in every boardroom in motorsport. He was perfectly dialed into Bernie's negotiating tactics, which were primarily cooked up to drive people nuts. The routine was essentially good cop, bad cop. While the urbane, conciliatory Mosley engaged with whoever was across the table, Ecclestone delighted in showing how quickly he could lose interest. In fact, Bernie would say so. Or he would stand up while people were talking to adjust picture frames on the wall. Or he would schedule flights perilously close to meetings in order to squeeze negotiations shut.

For much of the 1970s, negotiating was all he seemed to do. The battle for control of the sport dragged on for nearly a decade between Ecclestone's FOCA and the confusing collection of acronyms that made up motorsport's world governing body. At first it was known as the Commission Sportive Internationale, which ran competitions for the FIA from one of the ritziest addresses in the entire world of sports: Hôtel de Crillon, Place de la Concorde,

Paris. (It's as if Major League Baseball were run from a suite of rooms in the Plaza.)

Then in 1978, it became the Fédération Internationale du Sport Automobile (FISA), under the presidency of a French blow-hard whom Ecclestone came to despise. His name was Jean-Marie Balestre, but he behaved as though his real name were Louis followed by a Roman numeral. He was a martinet, a former go-karting administrator, and an erstwhile member of the Waffen-SS during the Second World War—a fact that Ecclestone never hesitated to bring up when things got tense.

"I have nothing against the English," Balestre lied in one television appearance. "Even if they did burn Joan of Arc at the stake."

Balestre's problem was that in the space of a few short years, Ecclestone had acquired a tremendous amount of influence with at least half the teams on his side. Try as he might, Balestre couldn't ignore him. The war that erupted between FOCA and FISA wasn't about any one specific thing other than control of the sport, though the two sides opened plenty of fronts. They argued about technical regulations, about circuits. And above all, they argued about cash.

"It's your friend, the little guy," one of FISA's stodgier blazers once told Mosley. "Each year, he wants more and more money and the organizers can't afford it."

Races were run under the constant threat of teams pulling out at the last minute, drivers going on strike, or FISA yanking away world championship points, which was no way to run a league. The NFL didn't schedule games wondering who might turn up. Wimbledon didn't threaten to lock players out of Centre Court. The uncertainty was exactly what anyone hoping to appeal to sponsors didn't need. At the turn of the decade, the situation had devolved so badly that the FOCA teams, which were still crying poor, drew up plans to break away entirely from FISA and become their own governing body called the World Federation of Motorsport (WFM). They proposed an eighteen-race schedule in fifteen countries featuring all of the most famous circuits, including

Monaco, Monza, and a new Grand Prix in New York. The series planned to launch in 1981.

The only issue was that none of those circuits were really prepared to defy FISA and come on board with WFM. Go ahead and join Bernie, the organization told them, but be prepared to lose the right to host F1 and any other motorsport competition at your circuit.

The situation had become untenable. What was supposed to be the first Grand Prix of the season in Argentina had already been postponed by three months from January because FISA could only guarantee that nine cars would turn up. Then the FOCA teams arrived in South Africa in February to race for no championship points without Ferrari, Alfa Romeo, or Renault.

So with one side prepared to pull the cars and the other threatening to pull the tracks, Ecclestone, Mosley, and Balestre returned to the negotiating table. And in early 1981, after talks in Italy and Paris, they reached a compromise. FISA would control all technical and sporting matters. And FOCA, with all of its financial power, was left in charge of promoting Formula 1, which is all Bernie really wanted anyway. Despite technically being a constructor as owner of the Brabham team, he could live without being involved in drawing up the next season's specifications. Right under the nose of the entire sport, Ecclestone had carved out the business of F1 for himself.

That March, Bernie and Balestre signed the peace treaty named, at Balestre's request, for FISA's home in Paris. They called it the Concorde Agreement.

THE FOCA-FISA DEAL BROUGHT FORMULA 1 A MODICUM OF peace. It also left Formula 1 right where Ecclestone wanted it.

Among the various concessions he'd secured for himself, Bernie had made sure that the deal placed the sport's broadcast rights under FOCA control. Considering what the British and European TV landscape for sports looked like in those days, no one realized that they had surrendered a potential cash cow.

After all, few broadcasters carried any races live yet. What little footage appeared on British television was shot during short sections of the race, flown back to the UK, and cut into high-light packages for news programs. And that's when British broad-casters aired them at all. In 1976, the BBC had refused to show F1 for much of the season, because it objected to the Surtees team's sponsorship deal with the London Rubber Company, which involved adorning the cars with the name of its most famous product: Durex condoms. The decision meant that the British public missed out on that season's pulsating duel between James Hunt and Niki Lauda.

That the system could be so backward was hardly a surprise. The BBC and its rival network ITV could barely agree to show all ninety minutes of soccer, Britain's national sport, let alone Grands Prix. As late as the mid-1980s, a dispute between the broadcasters and the professional soccer clubs would see all live football disappear from English television for the better part of a year. But Ecclestone knew that television remained the fu-ture. He'd spent time in the United States, where the National Football League was turning TV into its purpose for existing. Bernie had also crossed swords with a young American agent dabbling in motorsport named Mark McCormack, who'd spent the 1960s and 1970s revolutionizing golf and tennis by build-ing his clients into household television names, starting with Arnold Palmer.

Now, Concorde Agreement in hand, Ecclestone had the po-tential to do the same for Formula 1.

The most significant deal came almost immediately, ahead of the 1982 season, with a cumbersome, bureaucratic beast called the European Broadcasting Union (EBU). The umbrella group represented ninety-two public broadcasters across the continent. And Ecclestone convinced them that F1 racing was a product they desperately needed. There were just a few conditions. No longer could they cherry-pick the best races and ignore the rest—it was too easy to show the Monaco Grand Prix and ditch Sweden. If they wanted Formula 1, they would have to commit to airing every single Grand Prix in its entirety, whether it was at Silverstone or at Suzuka.

The money from the EBU wasn't much to write home about. It was believed to be in the very low seven figures. But that wasn't the cash Ecclestone was really after. By securing large chunks of airtime for races on television stations across Europe, he had manufactured hundreds of hours of new exposure for F1 sponsors. Everything from the logos on the cars to the boards by the side of the tracks was now guaranteed eyeballs. All of which meant that Bernie and the teams could extract more from the sponsors.

A decade before English soccer performed a similar trick, Formula 1 had a platform for a new era of commercialism in the sport. But this was only the beginning.

Ecclestone renewed the EBU deal once, only to decide by the end of the 1980s that it no longer made financial sense. Bernie had been working his used-car dealer magic on TV executives all over the continent and figured out that the once-critical EBU agreement could now be dispensed with like an old tire. If he could sell the TV rights separately in each country, there was a fortune to be made. Who needed one bulk deal when he could make dozens of smaller ones with all of his new friends in the broadcast business, from the director of sport at the BBC to a brash cable executive he'd met in Italy named Silvio Berlusconi. The EBU warned Ecclestone he would regret it. Ecclestone hardly knew the meaning of the word.

By 1990, F1 claimed a global audience for the season north of 1.2 billion viewers.

In order to keep growing that number, Ecclestone pursued his efforts to professionalize the series. He moved the start of every Grand Prix to 2 p.m. Europe time so that people knew where and when to find the races—early starts in America be damned. He also convinced broadcasters to build programming around the start and finish of the events to generate more screen time for the cars and circuits. All of which pushed the value of F1's TV rights for a five-year cycle to around $120 million.

Yet the most remarkable thing about that price, which dwarfed anything happening in any other European-based sport, wasn't just the size of it. It was exactly how much of that money found its way directly back to Ecclestone.

Under the 1987 Concorde Agreement, a renewal of the first deal, the FIA received 30 percent of the broadcast revenue, while the rest went to Ecclestone and the teams. But after the end of the EBU agreement, the FIA got spooked about the future of television, which suited Bernie perfectly. So at the suggestion of an Ecclestone ally—a business based in Ireland that handled hospitality and advertising at the circuits—the FIA traded away its 30 percent stake to the TV rights for a flat fee of $5.6 million in 1992, eventually rising to $9 million. By taking the guaranteed cash now, Balestre had no clue how much he was leaving on the table in the future.

In a lifetime of questionable decisions, this was surely his most expensive. Bernie had beaten him once and for all.

AS ECCLESTONE CORRALLED THE TEAMS, THE CIRCUITS, AND television rights—all while fighting the blazers in Paris—another force had emerged for him to deal with, straight out of thick, smoky air: the cigarette companies.

The tobacco industry had first entered Formula 1 after British Petroleum and Shell cut back on F1 sponsorship and Firestone decided to start charging the teams for tires. Imperial Tobacco was the first to stump up some cash to fund Colin Chapman's madness at Lotus in 1968 and paid him £85,000 a year to slap Gold Leaf decals on the cars. British American Tobacco followed close behind. But the game really changed when one of the biggest players in the business appeared on the scene in 1971 with Philip Morris International.

Burning sticks of tar and a sport full of greasy rags might not seem like the most natural match. But to Philip Morris it was perfect. The company had been looking for a way to sell its Marlboro cigarettes in Europe after their wild success in North America. The secret had been a pivot from marketing Marlboros as an elegant smoke for elegant ladies before the war to selling them as the only choice for a man's man in the 1950s. The aesthetic was pure cowboy—John Wayne types with rough hands and muddy jeans, who smelled of horses, the outdoors, and fine tobacco.

"You're talking about gentlemen who were loners," says Patrick Duffeler, the executive tasked with taking Marlboro across the pond and making it the world's No. 1 cigarette brand. "The cowboy is in the heart of every American."

The problem was that no one in Europe knew the first thing about the Wild West. They'd seen the movies, but this wasn't what little boys dreamed of in France or Germany or Italy. For Europe, Marlboro needed a different kind of cowboy.

It settled on Formula 1 drivers.

So Duffeler was dispatched back to Europe with half a million bucks of Marlboro money to spend on F1. Born in Belgium, educated in the United States, and a former executive of Eastman Kodak, he had the profile to move easily in motor racing circles full of jet-setting European aristocrats, playboy drivers, and tricky businessmen. The first part of the Philip Morris plan was to spend the cash on drivers who would wear the red-and-white logo on their cars, jackets, and fireproof suits. The other was to buy Marlboro as much press as possible by giving motor racing journalists what they wanted most in 1971: not scoops, but free food and booze in the company of pretty girls.

That approach didn't last long. Within a year Duffeler told his bosses in New York that they could be a little more strategic with all the cash they were throwing around.

"Forget all the expenses of feeding the press," he told them. "Let's put the money where it matters. We're going to spend money on the teams themselves."

Which one? Well, technically it didn't exist. Marlboro simply picked the drivers it liked in whichever cars they happened to drive and took over their sponsorship. Their star was the Brazilian Emerson Fittipaldi, who was so covered in logos that he showed up to races looking like a cigarette pack. But by calling its stable of sponsored drivers the Marlboro World Championship Team, Philip Morris had managed to get itself in the F1 game without ever building a car.

Philip Morris wasn't alone. Imperial and British American Tobacco were funneling more money into the sport than teams had ever seen and the arms race was on. Once F1 had its first taste, the sport was hooked. It was the beginning of a decades-long association that would see the tobacco industry pour more than $4.5 billion into Formula 1.

The one team that didn't immediately jump on the tobacco train was Ferrari. "My cars do not smoke," Enzo had told Duffeler. But it was just a matter of time before they picked up the habit. In the years to come, Marlboro would become part of the team's most iconic paint jobs. In the meantime, Ferrari thanked Duffeler for his approach with a gift of a large yellow ashtray with the *Cavallino* painted on the bottom. Except Duffeler, the man from Philip Morris, wasn't a smoker.

What the ashtray told him, though, was that every door in Formula 1 was now open to him. The tobacco influence was here to stay and Duffeler made it his mission to make Marlboro indispensable. One tactic was to involve Philip Morris in Formula 1 safety. At a time when drivers were routinely burnt alive in horrific crashes, Duffeler realized that a little tobacco cash could go a long way to improving the barriers and emergency run-off areas at circuits.

"Be known as the people who fund safety systems," he told his bosses, who were some of the same people trying to play down the connections between smoking and lung cancer. "We don't want to be known for killing people."

Soon, they were known for something far more surprising: Philip Morris was partially running the sport. The company's influence was so great that it found itself squarely in the middle of the fight between FOCA and the motorsport administrators, thanks to Duffeler's close ties to all the local promoters and circuits that he was paying to display Marlboro signage.

As ever in F1, the situation was complicated and full of acronyms. But it came down to Duffeler trying to unite the circuits against Ecclestone by forming a group called World Championship Racing. Marlboro wanted to work with the tracks and give them a larger say. Ecclestone wanted to squeeze them.

"Bernie's strategy was to tell every race organizer that they weren't important," Duffeler says. "He told them that he didn't really want more Grands Prix, the teams didn't want more, and they were the ones getting cut."

Yet Bernie's strategy was the one that worked. By pitting organizers against each other, he was able to create just enough doubt and dissension to break the union. In the mid-1970s, he crafted backdoor deals with individual circuits and outflanked Duffeler for good. Though there would be further problems down the road and several more years of FOCA-FISA sparring, Ecclestone had put tobacco in the place he saw fit for the sport. It wasn't a race organizer or a constructor. It was a money faucet. And as long as the cigarette companies were happy to play that role, then Bernie wasn't about to turn them away.

"Mr. Ecclestone's goal was very evident," Duffeler says. "Running the world of motor racing."

BY THE 1990S, IT HAD BECOME A LEGITIMATE QUESTION AROUND the sport: who owned Formula 1?

Bernie Ecclestone insisted there was only ever one answer. He controlled the television rights and was well on his way to controlling the commercial rights too. He also was in charge of negotiating with the circuits and local promoters. And in 1993, his buddy Max Mosley was installed as FIA president as Bernie himself became an FIA vice president in charge of promoting motorsport. As far as Ecclestone was concerned, all of that amounted to controlling Formula 1.

"Nobody was there trying to stop me doing something that I thought was good," he says.

A few had tried and failed without ever disturbing him. What all of his opponents had found was that the appearance of chaos around Ecclestone was in fact his greatest weapon. No one was ever quite sure where they stood. No one could sense when things were about to change. No one could even look anything up. The only complete and accurate accounting of F1's operations existed not in regulatory filings or tax returns, but inside Ecclestone's head. Only he grasped how everything was connected, where the weak points were, and how it all held together. Which was precisely how Bernie liked it.

"I carry out my business in a very unusual way," he said. "I don't like contracts. I like to be able to look someone in the eye and then shake them by the hand rather than do it the American way with ninety-two-page contracts that no one reads or understands. If I say I'll do something, then I'll do it; if I say I won't, then I won't. Surprisingly, people seem to like that."

But some people very much didn't, namely the F1 teams. Once Ecclestone sold off Brabham for £5 million in 1988, he became the sport's full-time promoter, unbothered by minor day-to-day Formula 1 concerns, such as drivers and cars. He could focus on what he truly cared about. All major decisions about where F1 raced and how it looked now went through Bernie—and so did most of the profits. In 1989, as some driver salaries reached the unprecedented heights of $6 million a year, Ecclestone was pocketing $1 million a race.

The hefty sum emerged after organizers in Phoenix, host of that year's US Grand Prix, revealed that they were on the hook for $3 million in fees: a third was paid for logistics, a third went toward prize money, and the final third, a cool million, went straight to Bernie. Presented with these facts at the time, Ecclestone was his usual unapologetic self.

"I'll take as much as I can get," he told reporters. "I don't get as much as I should, though."

This from the man who was about to become the highest-paid executive in Britain. In 1993, his combined salary from the two companies he used for selling television and sponsorship rights amounted to $44.5 million. The people responsible for actually racing Formula 1 cars could hardly wrap their heads around what was happening to their sport. Ecclestone had been one of them, just another constructor hoping to build a quick car and find a half-decent driver to sit in it. Now he was profiting from their work on a scale that no one who had seen F1 in the ragtag 1960s and '70s had ever imagined. After all of Ecclestone's administrative maneuvering at the turn of the 1990s, the teams found that they were receiving less than a quarter of Formula 1's total revenue.

"You've stolen F1 from the teams," the constructor Ken Tyrrell shouted at him.

Bernie disagreed. Ecclestone felt that the only reason they could still afford to put fuel in their cars and tires on their wheels was him. What wasn't up for debate was that by the early 1990s, Ecclestone had made Formula 1 more popular than it had ever been. The sport was reaching its technological peak, driven by the brains inside teams such as Williams and McLaren. And F1 was now broadcast in more than a hundred countries. It boasted ratings of two hundred million viewers per race.

Twenty years after he'd entered the sport as an outsider, Bernie was now pulling all the strings. Even if no one was quite sure who *owned* Formula 1, there was no longer any doubt that the man known around the paddock as "the Supremo" was in charge.

5

Lap of the Gods

IT WAS A SUN-SPLASHED SATURDAY afternoon in May and the final qualifying session for the 1988 Monaco Grand Prix was drifting toward its midway point. This was the sort of time in a race weekend when most spectators have only half an eye on the track, as they pick at a plate of petits-fours and wait for the real drama to unfold in the final moments of the hourlong session.

On this occasion, the pastries would have to wait. Instead it felt like every last person in the principality was transfixed by what was happening down on the circuit, where one man in a red-and-white McLaren and a bright yellow helmet was throwing a Formula 1 car around the narrow streets of Monte Carlo like no one had ever done before.

Ayrton Senna was completely out of his mind.

He'd only been out of the pit lane for ten minutes, but as soon as his MP4/4 car with the red No. 12 on the front hit the track, the twenty-eight-year-old Brazilian had begun to rip off laps at a furious pace. Everyone else had struggled to crack 1 minute 28 seconds during the course of the weekend. Senna, in his first complete lap, posted a 1:26.5. He followed that with a 1:25.6. Then came a 1:24.4. In the space of just six trips around the circuit, Senna had put three seconds between him and the rest of the field. Pole position was already safely in the bag. But it was no longer about that. He kept pushing, faster and faster, searching for the limit.

In one sense, this was nothing new. Senna, who drove with the conviction that his skill came directly from heaven, had always treated qualifying like a religious experience. From the day he entered Formula 1, his uncanny ability to wring every last ounce of performance out of a car over the course of a one-off lap had set him apart. And to this day, many of Senna's individual laps are recalled in hallowed terms by F1 fans, each one memorialized in a sort of shorthand. Portugal 1985, the one where he opened up a three-second lead on the opening lap. Monaco 1984, where he produced the fastest lap in the middle of a monsoon. Donington 1993, when he passed three world champions in the space of a single lap. Suzuka 1989, when he squeezed an Acura NSX to within an inch of its life in a promo spot for Honda while sporting white socks and brown leather loafers.

None of those came close to what was unfolding along the French Riviera on that afternoon in 1988. As he crossed the line to begin his ninth lap and blasted toward the Sainte Dévote corner, Senna was about to produce the single most incredible lap in Grand Prix racing history.

This was not something that was ever supposed to happen in Monte Carlo. No racetrack presents a more searching examination of a driver's precision, commitment, and cojones than the Circuit de Monaco. The streets are ludicrously cramped, the surface of the road is patchy, the corners are impossibly tight, and the run-off areas are a figment of your imagination. Most circuits are lined with grass verges or beveled curbs to accommodate tiny errors. Monaco is lined with concrete walls—and perched on the edge of a cliff. Threading the needle around the two miles and nineteen corners of its twisting streets is so grueling that Nelson Piquet, a three-time world champion and Senna's countryman, likened it to riding a bicycle around your living room.

Yet as Senna swept through Casino Square, fighting the wheel as he rounded a bend, powered down the hill, and shimmied through the Mirabeau hairpin at the end of the first sector, he was already four-tenths of a second quicker than his own fastest lap.

No one was more astounded by what they were witnessing than Senna's own team back in the McLaren garage. For one thing, they had ordered him to slow down and return to the pits two laps ago. By then, he was already assured of starting the race from the front of the grid. There was nothing to be gained from risking the car any further, the team told him. Senna disagreed.

"Please, let me," Senna responded over the radio. "I want to do it, for me."

That sort of unnecessary risk-taking was exactly the sort of thing that would normally have caused McLaren team boss Ron Dennis to experience what is known in Formula 1 as a complete shit fit. Docked in the harbor a few hundred yards away was an ocean liner he had personally hired for the weekend, packed with sponsors worth roughly $50 million a year to McLaren. The last thing Dennis needed was for Senna to ruin the car on some pointless crusade to prove something to himself. But even Dennis was too astounded by what he was watching to do anything about it. In the far corner of the garage stood Alain Prost, a double world champion and ostensibly the team's No. 1 driver, staring at the TV screen wearing a red flameproof race suit and the grave expression of a man beginning to suspect he might not be the team's No. 1 driver for much longer.

Out on the track, Senna snatched at the gearstick with his right hand, dropping all the way down to first to take the 30-mph Loews corner, one of forty-six gear shifts during each lap of the circuit. He nailed the blind right-hand turn into Portier, before slamming hard on the throttle, his head bobbing in the cockpit as he hit 170 mph in the long tunnel under the Loews Hotel.

The most remarkable thing about Senna's one-lap pace was that, by conventional standards, his driving technique was a total disaster. In a sport where the greats are defined by how they corner, Formula 1 drivers are taught to smoothly apply pressure to the throttle as they move through the bend and out of it, spreading the weight evenly through the front and rear tires and maximizing grip. That is not how Ayrton Senna did things. His violent, wheel-yanking, seat-of-the-pants style involved smashing the throttle on and off through the corner. Objectively, this was highly unwise. But behind all that rapid-fire pedal-stomping was a complex and counterintuitive technique that Senna developed to test whether his car was at the very edge of traction. The only way to truly know if a Formula 1 car is at the limit is to change a driver input and see what happens. Like turning the steering wheel, for instance. If the car's path tightens, then it wasn't at the peak of grip. But if you turn the wheel and it stays on the same path, you are at the limit—or possibly way past it. By constantly modulating his driver inputs, through a series of small, super-fast movements with the steering wheel and throttle, Senna could test the car's response in real time and keep it on a razor's edge.

You didn't need a series of small, super-fast inputs to work out that Senna was at the limit this time, though. As his McLaren swung through the swimming pool complex at the end of the second sector, he was now 1.1 seconds faster than anyone else, including himself. Everyone could sense that something magical was happening.

Except no one who wasn't there could actually see it.

The most incomprehensible part of Senna's perfect lap is that no footage of it exists. The onboard cameras that have become commonplace in Formula 1 today weren't introduced until the following season. And the director responsible for the TV cameras during the qualifying session for the 1988 Monaco Grand Prix elected to train them not on Senna—the man who had taken pole position at each of the previous two races that season—but on the Benetton car driven by Alessandro Nannini. (Alessandro who?)

The only people who got to witness the greatest qualifying lap of all time were the fans in attendance, the crew members in the pit lane, and one dashing Brazilian who was floating some thirty feet above the track. Because at the precise moment that he rounded the final bend of the track, Ayrton Senna was deep in the throes of an out-of-body experience.

"I was already on pole, and I was going faster and faster. One lap after the other, quicker and quicker and quicker," he told Canadian reporter Gerald Donaldson some time later. "Suddenly I was nearly two seconds faster than anybody else, including my teammate with the same car. And I suddenly realized that I was no longer driving the car consciously. I was kind of driving it by instinct, only I was in a different dimension."

By the time he snapped back to reality, Senna had crossed the line in a time of 1:23.998. It was not the fastest time anyone had ever posted at Monaco—Senna himself had gone quicker a year earlier when F1 cars ran more powerful engines—but the chasm between him and Prost, in first and second, yawned at 1.4 seconds. Gerhard Berger, who wasn't blessed with an MP4/4 to drive, was 2.6 seconds back in third. Even Senna couldn't quite process what had happened. "On that day, I said to myself, 'That was the maximum for me; no room for anything more,'" Senna admitted. "I never really reached that feeling again."

The following day, Senna produced another mind-boggling moment, only this one was captured on camera. After storming away at the start, Senna led the race by nearly a minute from Prost when he lost concentration, loosened his grip on the wheel, and crashed into a barrier twelve laps from the finish. Senna was so disconsolate that when he extricated himself from his busted-up car, he left the circuit, marched back to the luxury Houston Palace high-rise, and watched the rest of the race from his ninth-floor apartment. When it was all over, he took a long nap. Hours later, no one from McLaren had the faintest idea where Senna had gone.

For all of Senna's frustration, the outcome of the race was inconsequential. In less than 84 seconds that weekend, he had produced what may be the greatest Formula 1 lap of all time. It would soon become clear that he was also driving what may be the greatest Formula 1 car of all time. And by the end of the season, he would be crowned world champion for the first time.

In Ayrton Senna, Formula 1 was looking at its first global star.

IT SHOULD COME AS NO SURPRISE TO LEARN THAT AYRTON
Senna da Silva nailed his entrance to Formula 1 with exquisite
timing.

Just as the series was reaching new heights of popularity,
courtesy of Bernie Ecclestone's string of television deals, Senna
rocketed into F1 in the mid-1980s and radiated Hollywood quality. He had everything—the looks, the charisma, and the freakish
ability to jam a car to its upper limit and keep it there for longer
than anyone else. Senna thought that made him God's gift to motor
racing, but it really made him God's gift to Bernie. At a time when
Ecclestone was growing Formula 1 into a worldwide business, with
races that now stretched across five continents, Senna arrived as a
ready-made leading man.

There may be no driver before or since who possessed such international appeal. Much as Pelé had done decades earlier in international soccer, Senna was both a quintessentially Brazilian icon
and yet also a global one. Something about his speed and fearless-
ness behind the wheel transcended national boundaries. His yellow
helmet was as much a symbol of class and derring-do for a Formula
1 fan in São Paulo as it was for a young kid growing up in Stevenage.

Senna was idolized in Portugal, where he scored his first F1
victory in a torrential downpour. He was adored in Britain, where
he lifted a British team to new heights. But nowhere went as gaga
for a Brazilian folk hero as Japan, where his success piloting a
Honda-powered car was credited with giving legitimacy to the
Japanese auto industry right as the country surpassed the United
States to become the world's biggest car manufacturer. Quite apart
from anything else, this helped make Senna fabulously wealthy.

What set him apart was his ruthlessness. Inside the cockpit, away from the track, at the negotiating table, he was a totally
inflexible character, utterly convinced that only he could be
right. Between that and his ambition, Senna wasn't going to let
anything get in his way.

His first experience racing cars came at the age of twenty-one,
when he spent a season in Formula Ford in the UK. Senna won
twelve of the nineteen races, but his young wife, Lilian, struggled

to adapt to their new surroundings. Britain in the 1980s was like Brazil if someone had sucked all the Brazil out of it. It was gray, dreary, and rained constantly. The couple wound up returning home before the end of the season. When it was time for Senna to go back the following year, he knew that they had to find a solution to his wife's unhappiness or risk it becoming a distraction.

So he left her in Brazil and divorced her.

"If I was going to make it to Formula 1, I had to give it all my time and attention," Senna said. "I couldn't do that if I was married, so we parted."

He was just as cold in his business dealings. Just fifteen years after Niki Lauda had become the first driver to earn $1 million a season, Senna persuaded McLaren and Marlboro in 1993 to pay him $1 million *a race*—and then demanded the money be deposited into his bank account every Wednesday before a Grand Prix. On one occasion when the wire transfer failed to clear in time, Senna refused to leave his home in São Paulo to report for duty at the Portuguese Grand Prix. Only once the team had sent a fax confirming the amount had landed safely in his account did Senna finally board a plane for Portugal. He promptly flew to the wrong airport, arrived at the circuit midway through Friday practice, and immediately crashed.

It goes without saying that this sort of brinkmanship only endeared Senna to the sport's master negotiator. Despite their being decades apart, Senna struck up a close friendship with Bernie Ecclestone. Ayrton trusted Bernie, something few others in the entire sport ever dreamed of doing. And over long chats at Senna's sprawling home in Brazil, Bernie quickly understood that Ayrton had what it took to transform the sport. Senna had entered Formula 1 with everything required to become a champion. But there were still two things he needed to become a legend. One was a car that would allow him to express his genius. The other was a rival who would push him to go beyond it.

As luck would have it, one man would provide him with both of them. His name was Ron Dennis, and he was in the process of remaking Formula 1 in his own image as well.

ENZO FERRARI WAS DRAWN TO FORMULA 1 BY HIS LOVE OF speed. Colin Chapman saw it as a way to follow his creative mania. But Ron Dennis came to the sport with a very different plan in mind.

He saw an F1 team as a way to make money.

Even in a sport of permanent, self-imposed change, this re-thinking of the very purpose of running a Formula 1 team was so radical that it turned a ragtag racing team based in a dingy industrial park on the outskirts of London into a billion-dollar business empire.

Over three decades as team principal at McLaren, Dennis would introduce modern corporate practices to an arena that still operated as a cottage industry in order to build a blue-chip automotive conglomerate involved in sectors from electronics to corporate communications. Unburdened by history, McLaren had nothing to lose and everything to gain by doing things differently from every team that had come before it.

A high school dropout from Surrey, Dennis arrived in the sport at seventeen with no money, no formal qualifications, and no background in motor racing. Precisely how he managed to land a job as a mechanic remains completely bewildering. But what Dennis recognized earlier than anyone else was that the sport was at an inflection point. As F1 hurtled toward the 1980s, the introduction of turbo engines and carbon-fiber chassis was about to usher in a breathtakingly complex—and breathtakingly expensive—new era.

The most important race in Formula 1 was now the scramble to secure a financial advantage. And long before he oversaw the first of his record 138 Grand Prix victories as a team boss, that was one race Dennis understood how to win. "From a business point of view, losing money means your company is failing, whereas making money means your company is succeeding," he said of his philosophy. "And I want McLaren to succeed—always have, always will."

To do that, Dennis needed investors. In the years since Colin Chapman had slapped a logo for Gold Leaf tobacco on the side of his Lotus 49, Formula 1 had experienced a sponsorship boom, as companies fell over themselves to grab at the sport's irresistible mix of glamour, danger, and unbridled testosterone. Ron Dennis, a balding, straightlaced figure whose lips are set in a permanent sneer, embodied exactly none of those qualities. But no one in F1 history would prove so adept at harnessing the power of corporate partnerships to achieve their ambitions. "For me, image is key," Dennis says. "We can't win all the races, but we can always look the best."

Even before he got to Formula 1, his teams in the lower series of motorsport became the benchmark for presentation. Possessed with a fanatical attention to detail and an obsession with cleanliness, he instilled his punishingly high standards into every facet of the team. At the same time that Bernie Ecclestone was introducing his Brabham team to the groundbreaking concept of sweeping the garage floor, Dennis was bringing a signwriter to the circuit to touch up the paintwork on his cars between sessions and instructing the team's truck driver to park in the paddock so that the names on the tires all pointed in the same direction. Everything reflected his eye for detail and obsession with a professional approach. Even when the team won a race, Dennis demanded his employees maintain stoical composure in the pit lane.

"When you see a doctor delivering a baby you don't see him jumping up and down," he said.

To many inside motor racing, a sport that liked to think of itself as sexy and dangerous, Dennis's approach seemed clinical, bland, and nauseatingly corporate. But sponsors absolutely ate it up. And no sponsor in the 1970s was bigger than Philip Morris International, whose commitment to the sport had kept growing since the days of Patrick Duffeler. In Dennis, the tobacco maker found a man who understood corporate branding so instinctively he might have grown up on Madison Avenue. Together with John Hogan, the Marlboro marketing director, Dennis devised a document that became known as the "Book of Sponsorship," carving up a race car's bodywork for the first time into different sections, each with a different price tag, so that multiple sponsors could each have their logos emblazoned on the same car.

By 1979, the suits at Philip Morris were so dazzled by the performance of Dennis's Project Four Racing team in the lower series that when he came to them with a plan to enter Formula 1 with a revolutionary chassis made entirely out of carbon fiber, there was no question that they had to be involved. There was just one problem. Marlboro already had an exclusive commitment with the McLaren F1 team, a former champion now floundering at the back of the grid.

There was no getting out of the McLaren deal. Marlboro had invested way too much over the years to jump ship and start over. But letting Dennis take his carbon-fiber F1 project elsewhere wasn't an option either. While Hogan and his team at Philip Morris chewed over this dilemma, Dennis identified his opportunity. He'd delivered a lot of value for sponsors over the years. Now it was time for some payback.

One morning in the fall of 1980, Dennis rang Hogan at his London office. He had come up with a way around the impasse.

"How about if I buy McLaren?" Dennis blurted out.

Hogan and his team had talked through dozens of potential solutions in the previous days. They never once landed on this one. It was the equivalent of the local pharmacist suggesting he launch a takeover of Walgreens.

"How would you like to go about that?" Hogan asked.

"It's very easy," Dennis replied. "You would have to help me."

"How?" asked Hogan, unsure where all this was going. "How can we help?"

During his five decades in Formula 1, Ron Dennis has been branded abrasive, egotistical, aloof, and arrogant. But no one has ever accused him of subtlety.

"You tell them if they don't sell me 50 percent of the company," he answered, "then you won't sponsor them anymore."

Hogan didn't need much convincing. The newly rebranded McLaren International, backed by Marlboro and with Ron Dennis installed as team principal, was launched in the fall of 1980. It's no exaggeration to say that perhaps no one else could have pulled off such a maneuver. It relied on all the qualities he brought to the table—his vast ambition, his business creativity, his tenacity, his negotiating skill, and his remarkable clarity of vision. But more than anything, it came down to the fact that long before anyone else, Dennis recognized where the sport was headed. He saw how all the crucial elements of Formula 1—racing, technology, finance, branding, and the commercial possibilities—formed one great big virtuous circle. And at McLaren, he set about engineering that circle with the same zeal he once lavished on race cars as a trainee mechanic.

His attention to detail went into overdrive. The company headquarters in Woking were remodeled into a state-of-the art workspace that combined the decor of a five-star hotel with the clinical purposefulness of a hospital. Everything from the temperature (69.8 degrees) to the lighting to the humidity on the factory floor was precisely calibrated. Dennis even began pumping in different fragrances in his bid to create the optimal working environment. (The color scheme was one area where no experimentation was permitted. The McLaren offices were uniformly gray, a color that Dennis believed conveyed a sense of smartness and professionalism. "It's no accident that most suits are gray," he said.)

While other teams gradually came to see the importance of corporate partnerships to their bottom lines, Dennis was now exploiting them to accomplish specific strategic objectives. US aerospace giant Hercules agreed to construct his carbon-fiber monocoque in exchange for a small logo on the nose of the car. The engine that powered that car was supplied by a Saudi-born French entrepreneur named Mansour Ojjeh, whose investment firm Techniques d'Avant Garde was looking to invest in F1. Dennis persuaded Ojjeh to bankroll the entire design, development, and construction of a bespoke twin-turbo V6 engine he had commissioned from Porsche. In return, Ojjeh received a 60 percent stake in McLaren and naming rights to the TAG-Porsche engine.

Dennis had changed the calculus of what was required to compete in F1, upping the stakes to a level that most of his rivals were simply unable to match. They had mocked him for bringing his briefcase to the pits and for expressing himself in the manner of a business-studies handbook, talking about "global positioning," "vertical integration," and "layered management."

They weren't laughing when McLaren left them in the dust.

In his first season at McLaren, Dennis delivered the team's first victory in four years. In 1984, he clinched his first world title, as the TAG-Porsche engine propelled the MP4/2 to twelve wins from sixteen races. Dennis, thirty-seven, was the youngest team principal to win a championship since Colin Chapman. By this stage, however, he no longer thought of himself as merely the boss of an F1 team. Those people were hobbyists. In his mind, Dennis was something closer to a captain of industry.

He even began to dip his toe into different fields of business, teaming up with Mansour Ojjeh again in 1985 to acquire a luxury Swiss watchmaker. The brand would become known as TAG Heuer.

"If you asked me what I wanted written on my tombstone," Dennis would say later, "I'd say 'Ron Dennis, 1947 to so-and-so, one of the world's greatest entrepreneurs'—not anything to do with motor racing."

Dennis was still pretty good at the day job. By the late 1980s, McLaren's revolutionary, business-first approach had made it the dominant force on the track, the benchmark for technical excellence, and a pioneer in the field of sponsorship and marketing. It began to look as if there was nothing that Dennis couldn't do.

Which is exactly when he decided to put that theory to the test. For the 1988 season, Dennis executed a sequence of moves so radical that in any other sport they would have amounted to a full rebuild. In an aggressive bid to keep McLaren one step ahead of the field, he ditched the TAG-Porsche engine that had delivered twenty-six victories and three world titles. He assembled the most combustible driver pairing that Formula 1 has ever seen. And he hired a new chief designer, leaving the development of the team's car in the hands of an eccentric South African named Gordon Murray with a fondness for floral shorts, rock-and-roll music, and lightweight carbon fiber.

Individually, each of those moves was risky enough to derail his McLaren juggernaut. Together they added up to the most audacious moon shot the sport had ever seen.

By all rights it should have been an epic disaster. The decision to switch to Honda engines was so abrupt that four months before the start of the 1988 season, the MP4/4 didn't even exist. There was no prototype, no blueprint, no basic design concept—nothing. As if that weren't enough, many of the McLaren engineers had already come to resent the presence of Murray. There was no question that he was a brilliant designer, who pushed the boundaries of what was possible in Formula 1. But he was also a madcap personality who fit the austere corporate culture at McLaren like a pair of flip-flops with a three-piece suit.

And resentment was probably a generous description of the relationship that existed between the team's two drivers. Alain Prost, a bushy-haired Frenchman whose cerebral approach had earned him two world titles and the nickname "The Professor," was firmly established as the team's lead driver. For the 1988 season, he would be partnered by Ayrton Senna, a man who couldn't accept being the secondary anything.

Given all of this upheaval, it was hardly a shock when preseason rolled around and the MP4/4 was nowhere to be seen. The car still wasn't ready. McLaren just about got it together in time for the final day of testing at Imola.

Prost was first up that morning. As the Honda engine roared to life and the car lurched away, many of the McLaren engineers were quietly bracing for catastrophe. Ten minutes later, they couldn't believe their stopwatches. On his third lap, Prost was already 1.5 seconds quicker than the competition. On the fifth lap, the gap was up to 2 seconds.

"Everybody was looking at each other and asking themselves how this could be," said Dennis. "It was one of those surreal moments—almost like a movie going in slow motion."

When Prost returned to the pits, he gave nothing away. It was as if he'd just popped out for some milk. Understated as usual, he offered a few notes to the mechanics and suggested a minor tweak to the pull-rod that operated the front suspension.

When it was time for Senna to put the MP4/4 through its paces, he was a little more effusive. After a five-lap stint, the Brazilian returned to the garage. He removed his helmet, but remained motionless inside the car, an expression of incredulity stretched across his face. Finally, after thirty seconds, he broke the silence.

"This car," he whispered, "is going to be fucking quick."

Statistically, the MP4/4 remains the most dominant car of the sport's first seventy years. Prost and Senna won fifteen of the sixteen races, failing to achieve a clean sweep only when Senna clipped a straggler two laps from the finish at the Italian Grand Prix.

Midway through the season, Dennis received a call from Aleardo Buzzi, the president of Philip Morris International. For perhaps the only time in his life, Dennis had done something to upset one of his corporate partners. McLaren was winning too many races, Buzzi thundered. It was making the other Marlboro-sponsored teams look bad.

The only people who could prevent the McLaren drivers from winning, it turned out, were each other. Prost and Senna were teammates in name alone. Mixing like motor oil and Perrier, they were barely on speaking terms and informed Dennis from day one that they would never agree to any form of team orders. Prost would refuse to pull over for Senna and vice versa, even if it had been in the strategic interest of McLaren. Resentment between the pair had been festering from the first time Senna suited up in his McLaren overalls. For years, the team's policy had been that overalls were reserved exclusively for team sponsors. Drivers weren't permitted to wear the names of their personal sponsors on their team gear. Dennis thought it looked more professional and explained to his drivers that their generous retainers were paid to keep it that way and to compensate them for the loss of personal sponsorship. Yet when Senna arrived at the McLaren garage, the name "Banco Nacional" was emblazoned across the front of his team overalls. Dennis had made an exception—and Prost was furious. He had been turning away personal sponsors for years.

Once they made it onto the track, the relationship unraveled even faster. At Estoril in 1988, Senna had swerved toward Prost on the home straight and nearly shunted him into the pit wall. "I knew how much Senna wanted to be champion, but I hadn't realized he was prepared to die for it. If he wants it that badly, he can have it . . ."

Senna did want it that badly.

The season's result highlighted their essential differences. Senna—always on the limit, sometimes over it—took eight victories and three runner-up places. Prost, a master of consistency, had seven wins and seven seconds. Under that year's points system, Senna was world champion.

What began as a feud erupted into full-blown civil war for 1989. They reneged on a peace agreement in San Marino, crashed into each other in Japan, and turned any head-to-head into a game of high-speed chicken. The championship went to Prost. He promptly took his title and got the hell out of McLaren to Ferrari.

But he couldn't shake Senna. The 1990 season once again reached a climax at the penultimate race in Japan, where the two men started from the front row. On the first corner, Senna tried to pass Prost and knocked both himself and his only rival out of the race. The upshot was that Prost could no longer rack up enough points to overtake him in the standings. Senna had his second world title in the bag as he walked back to the pits.

"As I have said often since I met him first, he tries to represent to the world a man that he is not," Prost seethed afterward. "The world championship is for sport, not war—but he doesn't know it."

It would take more than McLaren-on-McLaren crime to slow the team down. For Dennis, seven drivers' titles in eight years were all the proof he needed that his uncompromising approach should be the only approach. In the space of a decade, he had utterly transformed a team that was languishing at the back of the grid and had shown the sport that individual championships could be won with innovative upgrades, but dynasties were built on much bigger bets. As he began to think about the next decade, he saw no reason why it shouldn't continue. He had the biggest budget. He had the fastest car on the grid. And he had the fastest driver in the sport. A dominant future stretched ahead of him like a wide-open straightaway.

There was just one eventuality that Ron Dennis hadn't accounted for. Some fifty miles up the road, another Formula 1 team had been doing some of its own thinking about the future, and it was about to come crashing right into the present in the form of a race car that was so technologically advanced it would upend the delicate balance of man and machine and challenge the very nature of F1 as a fundamentally human endeavor.

Over at Williams, it barely mattered who was driving the thing at all.

THE HEADQUARTERS OF FRANK WILLIAMS'S LITTLE ENTER-prise in Oxfordshire was not the most obvious place to go looking for the next quantum leap in Formula 1 in late 1991—and not just because this was a team that started life operating out of an abandoned carpet factory.

Only a few years earlier, Williams had been involved in a car accident that left him confined to a wheelchair. Racing down a country lane in the south of France in the spring of 1986, he lost control of his rental car, which careened off the road, hit a wall, and fell eight feet into a plowed field. The car landed upside down, trapping Williams between his seat and the crushed roof, breaking his neck. In the days that followed, doctors in Marseille declared him clinically dead three times.

Williams survived, but he would never walk again. The accident rendered him a quadriplegic, paralyzed from the chest down. He was forty-three, married, with three children.

Had Williams quietly folded at that point, no one would've had the faintest inkling that the sport of Formula 1 had been robbed of a burgeoning powerhouse. Because before he built his team into a nine-time world champion, before the road accident that changed his life, and long before he became a knight of the British Empire for his services to motorsport, Frank Williams was regarded as a complete joke.

His story reads no differently than those of the many other young men who were drawn to motor racing in postwar Britain. Born and raised in the northeast of England, Williams grew up dreaming of driving race cars, not building them. But he had neither the natural talent nor the substantial fortune required to support a racing career. So he began running a team instead, selling spare parts on the side to finance the thing. In 1969, with enough money to buy a secondhand Brabham chassis and a couple of Ford Cosworth engines, he entered Frank Williams Racing Cars into Formula 1. The team motor home was a family caravan.

The only people interested in what was going on inside Williams's garage were his creditors. Frank was forever on the brink of bankruptcy. At one point, he took to running his business from a phone booth around the corner after his line was disconnected over another unpaid bill. In the pit lane, he was nicknamed "Wanker Williams" and dismissed as a start-line specialist, willing to put any driver behind the wheel of his car for a spare dime.

Even when Williams found an investor with deep pockets and a desire to empty them into a Formula 1 team, it somehow went wrong. In April 1975, Frank was introduced to a brash Canadian oilman named Walter Wolf, who had recently bought the last Lamborghini Miura ever to roll out of the Italian factory—then promptly decided it wasn't fast enough and gave it to his wife for the weekly grocery run. The sort of breakneck thrills Wolf was searching for could only be found in F1. So when Williams explained that the finances at his team were a little tight and he was sorely in need of a new engine, Wolf agreed to buy him one.

A few weeks later, Williams was back for another. "By the end of the 1975 season," Wolf recalled, "I had bought him eleven engines."

For the 1976 season, the two men came to a simpler arrangement. In exchange for wiping out Williams's debts, which had now swelled to £140,000, Wolf acquired a 60 percent stake in his Formula 1 team. That should have solved Frank Williams's troubles, except he found that Wolf's new setup came with a problem: there was no role in the team for Frank Williams. He had effectively been replaced as boss and reduced to a glorified gofer. "I sold him 60 percent—I should've made it 49 percent, but he was a much better businessman than I was," Williams said. "Very tough guy."

Williams resolved to start over. Despite all those years of struggle, his childlike enthusiasm for motor racing never waned. Frank remained convinced that success was always waiting just around the next chicane. And this time, he was actually right.

The change in fortune wasn't the result of any master plan or a sudden epiphany about a virtuous circle. It turned out that building an F1 winner didn't need to be any harder than finding a brilliant technical director and a sponsor with bags of money.

He didn't have to look far for the technical director. Williams had a hunch that one of the young engineers back at Wolf Racing would fit the bill. Patrick Head struck him as a man who could go places. And as luck would have it, finding the right sponsor didn't require much digging either. In a chance encounter at a London ad agency in 1977, Frank Williams ran into an executive looking for a way to raise his employer's profile in Europe.

Without knowing it at the time, Williams had met someone who would make an oilman like Walter Wolf look like he was sitting on a dry hole. In exchange for a logo on the back of the car and a promise that his drivers would abstain from drinking champagne on the podium, Williams received a check for £100,000 and an ongoing financial commitment from the kingdom of Saudi Arabia. Within two years, the Albilad-Saudia Racing

Team Williams was Formula 1 world champion, flying the green Saudi flag next to Britain's Union Jack.

That success was largely down to Head. The son of a brigadier general in the British Army, Head was stout and square-jawed and had inherited his father's gruff voice, purposeful gait, and penchant for barking orders at people. He proved to be a practical, no-nonsense designer with a gift for taking the best innovations in the paddock and wrapping them into a cohesive, functional package. His first championship car at Williams was essentially a refined version of the ground-effects Lotus 79. (Refining ideas held next to no interest for Colin Chapman, whose mind had already moved on to hunting for the next radical breakthrough.)

Since Frank's accident, Head had effectively taken control of the day-to-day running of the team and had hired some of the brightest minds in the sport, including a promising young aerodynamicist named Adrian Newey. But there was little to suggest that Williams was about to produce a revolution that would knock Ron Dennis and McLaren off the top step of the podium. Head's cars were famed for their reliability, drivability, and dependability. They weren't game-changers.

But there was another of Chapman's ideas that had been lodged in Patrick Head's brain for the best part of a decade—and no matter how hard he tried, he couldn't shake it. Over the years, some of F1's most innovative thinkers had tried to make it work without success. But in 1991, Head began to believe it just might be possible. And this one wasn't a loophole so much as a hack.

Patrick Head was about to break Formula 1.

The concept he had been chewing over was something known as active-ride suspension, and it sought to solve a problem that had dogged F1 engineers ever since they began messing about with aerodynamics. Early on, they realized that one of the keys to maximizing downforce was keeping the floor of the car at the same constant height off the ground. They also realized that keeping a race car at a constant height off the ground as it blasts its way around a bumpy, winding racetrack at 200 mph was completely preposterous.

No matter how much you stiffened the suspension, there was no way a coiled spring could absorb all the shocks and forces exerted on an F1 car over the course of a two-hour Grand Prix without bouncing wildly up and down or breaking into smithereens.

Chapman's Lotus was the first team to conceive of another way. By using hydraulics in place of springs, it was possible to selectively tighten the suspension on each individual wheel in order to keep the car level. Williams began working on its own version in 1985, modifying a system originally devised to keep ambulances steady while rushing patients to the emergency room. Both those early efforts yielded the same result. Active suspension made for a smoother ride, but not a faster one. "It's amazing—it rides just like a Cadillac," Nelson Piquet told the Williams engineers after a test drive. "There's just one problem—it handles like a Cadillac as well."

The issue, Head realized, was that active suspension was still too reactive. It could respond to bumps and corners and mitigate their impact on the car's ride height, but what he needed was something that could anticipate those variables in advance. Something that could calculate the car's speed and the contour of the road ahead and tweak the suspension on the fly—rigid one second, squishy the next. Not just from race to race, or even lap to lap, but from corner to corner.

Back in 1985, all of that seemed to belong to some distant future. But in just six years, the distant future screeched into view. The computing revolution was about to transform the automotive industry. And nowhere was better positioned to take advantage of the nascent power of the microchip than Williams Grand Prix Engineering.

Head had built Williams into a huge factory operation, one of the few outfits in the sport capable of engineering pretty much every part of an F1 car under its own roof. A team without a road car business that had begun life as a collection of spare parts had wagered its future on remaking itself into its own in-house manufacturer. That meant producing not just the engine, the exhaust, the chassis and gearbox, but also the electronics.

The last of those would prove crucial. In 1987, Williams hired a young Cambridge graduate named Paddy Lowe and set him to work on an electronic solution to his active suspension problem. In four years, Lowe created Formula 1's first functional onboard computer.

If that sounds unremarkable in today's era of computer-controlled cars, it's worth remembering that in 1991, most consumer PCs were as bulky and cumbersome as a carry-on suitcase. The device that Lowe bolted onto the Williams chassis was the size of a small paperback. It weighed just over two pounds.

No high-speed automobile had ever worked with silicon in this way. Built from scratch inside the Williams factory, from hardware to software, the computer could reprogram the suspension remotely from the garage, raising and lowering the car in such subtle and rapid fashion that it remained perfectly level at all times. It was worth over a second per lap.

Head and his team realized they were just scratching the surface. They began to conceive of a car that didn't rely on the reaction speed and reflexes of the guy behind the wheel but was governed by software—where not just the suspension but the engine revs and the gear shifts were all controlled by a line of code and could be tuned from a PC in the pits.

In short, they began building a 200 mph supercomputer.

When it pulled up to the start line for the 1992 season, the Williams FW14B was the most complex race car F1 had ever seen. It seamlessly blended advanced aerodynamics with groundbreaking technology and mechanical reliability. The onboard computer managed not just the active suspension, but also a semiautomatic gearbox, traction control, onboard telemetry, and electronic data logging. Every part of the FW14B was at the bleeding edge.

Well, almost every part.

The man Williams handpicked to pilot this groundbreaking piece of machinery was something of a throwback. In fact, he was the oldest driver in the sport—a middle-aged father of three with a bushy mustache and the beginnings of a dad bod. His name was Nigel Mansell, and at the age of thirty-nine, his most notable distinction was driving the most F1 races without winning a championship.

That was about to change. Because it turns out the FW14B didn't need the speediest or most skillful driver on the grid. All those electronic gizmos took care of that. What it did need was a driver with the courage to take a corner at 130 mph while waiting for the active suspension to kick in—and the brute strength to keep the car on the track once it did.

Mansell was the perfect match. His heavy physique meant he had the upper body strength to withstand the extra downforce that came from cornering at such high speeds, while his record of 32 crashes in 172 races was evidence of a driver unafraid to push his car to the absolute limit.

"The combination of Nigel's balls and that '92 car was unbeatable," Paddy Lowe told *Autosport* magazine.

From the moment the lights went out at the start of the season, Mansell hit the throttle and barely let up. He reeled off victories in each of the first five races as the FW14B made the rest of the field look like they were pedaling push bikes. Mansell duly clinched his first title with one-third of the season still left to run.

If Mansell thought his status as a newly minted world champion had earned him a fat new contract, Frank Williams had other ideas. "Drivers are like lightbulbs," he was fond of saying. "You just plug them in." Hours after Mansell was crowned at the Hungarian Grand Prix, he announced he was quitting F1 after negotiations over a new deal for 1993 fell apart. (The final sticking point reportedly came down to the number of Mars bars the team would supply to Mansell's hotel room.)

Williams didn't miss a beat. It simply screwed in a replacement, persuading Alain Prost to suit up again at the age of thirty-eight. He'd been turfed out of Ferrari after comparing his car to a truck and chose to retire. Prost was given the keys to the FW15C, a car that blurred the line between man and machine even further. Head and his crew had added power steering and antilock brakes. It was an FW14B, only faster, better, and smarter.

Prost stomped the field on his way to winning the world title, practically on autopilot. By the end of the season, Damon Hill—an erstwhile motorbike racer who didn't begin Grand Prix racing until his early thirties—was no slower in the car than Prost, a four-time world champion.

This state of affairs so thoroughly offended Prost that he retired from F1 at the end of the season, this time for good. That alone was not unduly concerning for Williams. At this point, all it needed was a driver who knew how to buckle his seatbelt and spray champagne from the top of the podium. Except there was one person even more offended than a Frenchman who'd had his nose put out of joint. The man who ran Formula 1 had decided enough was enough.

Watching Williams turn every Grand Prix into a procession was bad enough for Bernie Ecclestone. But the prospect of a Formula 1 driver becoming no more than a glorified passenger was too much. In early 1993, the sport's governing body passed an edict banning electronic driver aids for the following season.

Williams appealed the decision, of course. But Ecclestone was unmoved. "The engineers are always unhappy if you take away their toys," he said. Just like that, active suspension, traction control, and antilock brakes were all outlawed.

The last great technical leap in Formula 1 had been legislated out of existence.

ACTUALLY, THERE WAS ONE PERSON WHO HAD FOUND WILliams's crushing superiority even more disagreeable than Bernie Ecclestone.

Ayrton Senna had spent two seasons watching their blue-and-white cars disappear into the distance and he'd had quite enough. Running half a lap behind Nigel Mansell for most of the 1992 season was one thing. Watching Alain Prost, of all people, pitch up at Williams and subject him to more of the same for a second consecutive season was frankly unbearable. Senna was so incensed that he offered to drive for Williams for free. Prost, who had already signed his contract with Williams for 1993, considered the possibility of sharing a garage with his former McLaren teammate for approximately one nanosecond before delivering an unequivocal *non*.

"There was only one thing I asked," Prost said later of his contract talks with Frank Williams. "You give me the money you want. I don't want to be the No. 1 in the team. But all I ask you is—I can't be a teammate again with Ayrton."

Senna wouldn't have to wait long to get the drive he wanted most. Prost announced his retirement after clinching his fourth title at the 1993 Portuguese Grand Prix, opening up a seat for a lead driver on the sport's leading team. Fifteen days later, Senna inked a deal to drive for Williams in 1994. "It is a dream come true," he said. "I'm really looking forward to driving a Williams-Renault in what I consider the beginning of a new life in motor racing for me."

That new life was tragically cut short before Senna ever had the chance to complete a season in his dream car. Instead of putting the spotlight back on Formula 1's leading man, 1994 turned into one of the darkest years in the sport's history. On May 1, at the San Marino Grand Prix in Imola, Senna lost control of his Williams exiting the flat-out Tamburello corner and struck a concrete barrier. The car's onboard telemetry said Senna was traveling at 193 mph when he lost control. Though the deep black tire marks on the track testified to violent braking in the split second Senna realized he was in trouble, his FW16 made impact at a speed of roughly 140 mph. The right front wheel flew back toward the cockpit, striking Senna's helmet.

"Tamburello was always a corner where your heart was in your throat," says former world champion Keke Rosberg. "Because you knew, if you went off there, that how you hit the wall was simply a matter of luck, good or bad."

At 6:40 p.m. local time, just over two hours after the race had ended, Senna was pronounced dead from severe head injuries. He was thirty-four. Officials who examined the wreckage found inside his car a furled Austrian flag that, had he won, Senna planned to raise in honor of Roland Ratzenberger, a driver from Salzburg who had been killed in a high-speed crash during qualifying just twenty-four hours earlier.

Grief for the Brazilian, struck down in his prime, spread worldwide. The government of Brazil declared three days of public mourning capped by a state funeral, which drew three million people to the streets of São Paulo. Alain Prost served as one of the pallbearers.

The death of the sport's most popular driver changed Formula 1 in ways both predictable and highly perplexing. The FIA immediately implemented a raft of measures aimed at improving driver safety, including changes to many circuits, mandatory testing of tire barriers, and stricter standards for helmet design. In the three decades since Imola in 1994, there has not been a single fatality during an F1 race.

More puzzling is the notion advanced by Bernie Ecclestone that Senna's death actually grew the sport's global audience, that the wall-to-wall coverage of his crash around the world brought viewers to Formula 1 in greater numbers than had ever been seen before.

"His death got so much publicity worldwide—again, free of charge," Ecclestone says. "It's only when something like that happens that you realize these things."

It may be a callous and chillingly transactional way of looking at the death of an icon. But Ecclestone's observation also contains a kernel of truth. Before Imola in 1994, it had been twelve years since the last fatality in a Formula 1 race, when the Italian driver Riccardo Paletti plowed into the back of a stationary car at the start of the 1982 Canadian Grand Prix.

Safety improvements meant the specter of death no longer hung over every race, and while no one disputed that those advances—many of which Ecclestone had ushered in himself— were a good thing, there are those who felt they had also robbed the sport of some of its essence. Grand Prix racing became popular in the 1950s and '60s because it combined the glamour of young men piloting fast cars around winding tracks in exotic locales with the danger of young men piloting fast cars around blind corners in deep forests. No one wanted to see drivers get killed. But when that element of danger was lost, so was a tiny sliver of intrigue. A driver's primary concern was fuel loads and tire compounds, not whether he'd survive to see the finish line.

To what extent Senna's death reawakened people to the essential perils of racing a car at 200 mph remains unquantifiable. What definitely was quantifiable was its impact on the size of the audience. By the mid-1990s, more people were watching Formula 1 than ever before, with close to two billion viewers a season. And when they tuned in, they were missing a star, but about to witness the rise of a dynasty.

6

Schumi

ENZO HAD BEEN GONE FOR nearly three years when the call came for Luca di Montezemolo to return to Ferrari. Back in the 1970s, that's where a much younger Luca had begun his career as an assistant to the Old Man, been promoted to team manager, and brought order and glory back to the Scuderia.

Now, in late 1991, they needed him to make magic again.

The inheritance that Enzo had left behind was less Tuscan villa and more like a collection of chipped tea sets and moth-chewed tuxedos. The Scuderia hadn't won a Grand Prix in over a year, or a drivers' championship since 1979. Unable to build a competitive machine, it finished a distant third in the 1991 constructors' standings, humiliated by the more agile McLaren and Williams, who had better technology, better engineers, and better drivers. Ferrari was running on horsepower and history alone. Just as troubling was the realization that the road car business was losing its shine too, as Ferraris became little more than gaudy accessories. So as Montezemolo looked around the place he used to know, he could tell that Enzo's fusty estate was falling into disrepair. What Ferrari required, he says, was "to open the window, to have fresh air entering the room."

New ideas had rarely been welcome under Enzo. But during his years away from Maranello, Montezemolo had been on a grand tour of other Italian sporting projects. Under the Agnelli family, he had served as a vice president of their soccer club, Juventus. He also oversaw the country's entry on the America's Cup scene with a yacht called *Azzurra* before skippering the organization of the 1990 World Cup in Italy. His monthlong spectacle included fifty-two matches across twelve cities and even made room for the first ever joint performance of the Three Tenors in Rome.

"The only terrible thing was the final match between Germany and Argentina," he says of the dull German victory. "One of the worst matches I've ever seen in my life."

He soon discovered that the only thing more stultifying than *1–0 Deutschland* was what went on in the Ferrari factory, where it seemed nothing ever changed. Years of inertia had seen an entire generation of Formula 1 development pass the team by. Instead of living at the cutting edge, Ferrari treated aerodynamics, electronics, and the mastery of carbon fiber as secondary concerns. The Scuderia tradition, after all, was to manufacture beasts of mechanical engineering, not computer-guided missiles on wheels.

"Engine, power, gearbox," Montezemolo says. These were Enzo's Holy Trinity.

Rescuing Ferrari would require some light blasphemy. That, and some $200 million.

The most important upgrade was the construction of a state-of-the-art wind tunnel. The Brits, who had spent decades fiddling with scale models and having their hair blown off, already knew that any competitive F1 team needed its own aero testing facility. Ferrari had simply never bothered. But in his first years back, Montezemolo ordered up a hulking wind tunnel, designed by the Italian architect Renzo Piano, to accommodate both 50 percent scale models of F1 cars and the full-size cars themselves. Equipped with a fifteen-foot fan, the tunnel required as much power to run as the lights of two thousand apartments. Ferrari spared no expense and built its own electrical substation to supply the juice.

Later, when the time came to reimagine the rest of the factory, Montezemolo went just as big. And this time, he employed the award-winning French architect Jean Nouvel to design the spiritual home of Italian automotive style. (Montezemolo's attention to aesthetics and the emotional connection between consumer and product would never leave him. Later in life, as the founder of an Italian high-speed rail company, he would boast that his trains had "30 percent larger windows than our competitors.")

Inside those buildings, Montezemolo went about streamlining every management process and took stock of Ferrari's F1 suppliers. He needed more from all of them, he said. And to make all of those partners feel more special, Montezemolo created Ferrari's annual supplier awards dinner, plying them with pasta and booze until they felt the Scuderia mystique rub off on them. What happened at Ferrari wasn't mere assembly, he explained, not like the *garagistas*. It was pure Italian craftsmanship from a small Italian village.

Montezemolo applied the same diligence and mythmaking to the road car business, because at Ferrari, racing and road cars were more intimately linked than anywhere else. One could not be successful without the other. Agnelli had charged him with turning it around, because, let's face it, the Prancing Horse wasn't the show pony it used to be. Models like the 348 and the Testarossa were dated and tired and Ferrari had become complacent. Worse still, it had become too accessible. Montezemolo's first act was to reduce the number of units produced by the company immediately. In the space of just three years, he nearly halved the number of cars rolling out of Maranello to some 2,300 in 1993 from 4,300 in 1990. Scarcity and quality were essential. Not only did Montezemolo decide that there would be fewer new Ferraris for sale, but he decreed the company should take longer to deliver them—on purpose, specifically to make the product seem harder to snag. The Ferrari waiting list, now around eighteen months long, was a Montezemolo invention.

"It's better to sell one car less than the demand to protect the exclusivity of the brand," he says, describing a core tenet of the luxury world. "You have to *desire* a Ferrari."

Desiring a job at Ferrari was another matter. Who would want to work for a third-place team that was years behind its more sophisticated rivals? Even the people who already worked there didn't especially want to work *there*. Its chief designer John Barnard, a former star under Ron Dennis at McLaren, refused to move to Maranello when he joined in the late 1980s and was sketching Ferraris remotely from his home in Surrey.

That's not to say Barnard wasn't doing innovative work, despite some of the results. He pioneered the semiautomatic gearbox for Ferrari, allowing drivers to change gears by flicking paddles behind the steering wheel instead of taking one hand off to operate the gearstick. But the situation of not having the man designing Ferrari F1 cars anywhere near Italy just didn't sit well with Montezemolo.

The question of how Italian Ferrari needed to be had been a recurring theme in the team's history ever since Enzo found himself on the wrong side of the Vatican. Plenty of outsiders believed that more Italianness only meant more chaos.

"You've got a whole bunch of Italians who all believe that they know the right thing to do," Bernie Ecclestone told Piero Ferrari for years. "All arguing amongst each other."

Montezemolo knew that the push and pull on the team's national identity was a more subtle balance. He wouldn't compromise on where his directors did their jobs—that needed to happen inside the walls of Maranello, living and breathing the history of the team. But he wasn't so beholden to Ferrari tradition when it came to where those directors came from. Showing the flexibility required to actually compete in modern Formula 1, he hired a Frenchman to run the team.

Jean Todt had precisely zero background in F1 when he accepted Montezemolo's offer to lead the Scuderia in 1993. The son of Polish Jews who'd fled eastern Europe and settled in France, he was more of a rally car man. Todt got his motor racing kicks navigating across wild terrain in two-man cars for hundreds of miles at a time. Closed circuits and open wheels were a different world entirely. Yet Montezemolo saw plenty to like. Todt had two qualities that he was looking for in a Ferrari team principal: he was a company man and he was a political animal. (Montezemolo chose to overlook the faux pas of Todt turning up to his first meeting in a Mercedes.)

Todt had spent virtually his entire career at Peugeot, climbing the ranks at the family-run manufacturer in the east of France until he led its sports division. He oversaw victories in Paris–Dakar, the World Rally Championship, and at Le Mans. Since 1975, he'd also been involved with motorsport's governing body, learning its inner workings as a representative for rally drivers. Todt spoke the FIA's language—and not just because he was French. Soon, he'd be speaking fluent Italian as well.

Todt joined just in time to assist with the earliest stages of the Ferrari turnaround. There were still no wins for the Scuderia in 1993, as it slumped into fourth place, but Montezemolo drew hope from a couple of podium finishes in Spain and Canada.

"The first drops of rain will begin to fall in the desert," he told reporters.

A full-blown shower would have to wait until the following season. After fifty-eight Grands Prix without a Ferrari victory, the streak ended with Gerhard Berger crossing the line first in Germany that July. Never mind that only eight cars managed to finish the race. Ferrari fans were overjoyed to have a winner again. Except this was no time to get carried away. Progress was achingly slow. Though the Scuderia earned podium finishes in more than half the Grands Prix in 1995, Montezemolo and Todt remained stuck on a single race victory. For all of the steps they took on the car, the factory, and the team's personnel, they knew they needed one more major upgrade. And this was one part they couldn't manufacture themselves. Early in the season, Todt pulled out the notebook that might as well have been a window into his brain. He added one more item to his to-do list:

"Driver Problems for 1996."

THE DRIVER SOLUTION FOR 1996 COULDN'T HAVE BEEN MORE obvious. If Ferrari wanted to be a world champion constructor, only a world champion driver would do. On a yacht belonging to Fiat president Gianni Agnelli at Monaco in 1995, Todt and Montezemolo welcomed aboard a tall, square-jawed German to make their pitch.

Michael Schumacher wasn't yet the greatest of all time, but he was on his way.

Anyone who had been around for Schumacher's first week in Formula 1 had known this for a while. Schumacher's debut came out of the blue in the summer of 1991. The Jordan team was preparing for the Belgian Grand Prix when it suddenly lost one of its drivers, because he'd assaulted a London cab driver. So Jordan scrambled to replace one maniac with another. The week before the race, the team organized a brief testing session for a hyper-competitive young German who was dominating the Formula 3 series. Those few laps in the English countryside were enough to convince Jordan that the twenty-two-year-old Schumacher was ready. He made the point emphatically once he got to Spa by qualifying in seventh that Saturday. His Sunday, however, was less memorable. Schumacher's clutch burnt out almost immediately and the car couldn't complete the first lap.

But the wheels were already in motion to put Schumacher in a better car as soon as possible—and even Bernie Ecclestone was in on it. That better car turned out to be on the Benetton team, run by an Italian businessman and convicted fraudster named Flavio Briatore. That he would later host the Italian version of *The Apprentice* tells you everything you need to know about Flavio Briatore.

The team had only been around since 1985, when fashion entrepreneur Luciano Benetton decided that the best way to sell more of his colorful sweatshirts was to become a major player in Formula 1. But unlike some of the more esoteric branding exercises that would come in later years—see: the outfit built by an energy drink empire—Benetton wasn't all that out of the ordinary. The team hired (and often fired) established F1 minds to build its cars and recruited future stars to put behind the wheel.

The only truly unusual employment decision was the man in charge. Flavio hadn't come from motor racing at all. Rather, the silver-haired Piemontese with a taste for the high life and dating supermodels had been Benetton's commercial director. He was flanked by Tom Walkinshaw, a Scottish former driver who brought some racing know-how, forming a strangely functional odd couple. By 1992, the Italian team with cars painted baby blue was established enough to beat the Italian team with cars painted bright red.

"People maybe didn't take us seriously, thinking we could only make shirts," Briatore said that season. "But now they appreciate the reality of the situation."

This new reality was that Briatore and Walkinshaw had enough sense to employ two of the most coldly rational people on the paddock to handle their cars: Michael Schumacher to drive them, and an owlish Brit named Ross Brawn to build them. Walkinshaw had known Brawn during their days in the World Sportscar Championship, where Brawn designed a title-winning car for Jaguar. The plan was for him to do the same in Formula 1.

Their timing was accidentally brilliant. The 1994 season found the sport in a rare spell of regulatory chaos. The FIA had outlawed the Williams computer aids and, as a way to appease Ferrari's lust for horsepower, kept the rules on engines surprisingly broad. So while Williams and McLaren ran V10s, Ferrari was allowed to keep developing its erratic V12s. Benetton, meanwhile, stuck with the smaller Ford V8s. But what the team lacked in brute force, it made up for in racecraft.

Ross Brawn was about to find a devastating speed advantage in the only moments the car stood still: pit stops.

For most of the sport's history, those trips through the pit lane had been seen as an annoying yet essential waste of time for a change of tires and, until 1983, a tankful of gas. Brawn saw them as an opportunity, especially once refueling returned in 1994. With a revolutionary insight that would eventually become the only way things were done in F1, he broke down races into a series of stints between stops. By quantifying precisely how much fuel the car would burn and how much rubber the tires would destroy, he turned the Benetton into the most efficient machine on the circuit.

Because the approach was so technical, it required a driver of consummate skill and intelligence to pull it off. Luckily, the man in the cockpit was a master of car control. Schumacher was peerless at conserving fuel when necessary and then going hell-for-leather for half a dozen laps.

"I just liked to compete," he said, "and usually to win."

So when he was at the wheel, Schumacher didn't ask himself many existential questions. He wasn't an engineer manqué like Alain Prost. Nor did he believe that entering an F1 cockpit put him on a divine plane like Ayrton Senna. Schumacher just raced whichever car was in front of him—and either put it behind him, or put it in the wall. Everything about this sport was just another problem to be solved.

When his childhood go-kart wasn't handling properly, he taught himself to be his own mechanic. When his F1 engineers wanted to try out different setups, he'd be at the test track at dawn to work through them. And when the problem was his Williams rival Damon Hill, well, Schumacher dealt with Damon Hill too.

The incident came at the end of the 1994 season. Schumacher had flown to Adelaide, Australia, for the final race on the calendar, leading Hill in the drivers' championship by just one point and chasing the first title for a V8 engine in over a decade. After thirty-five laps of the Grand Prix, the struggle for the title couldn't have been closer. Schumacher was in first place. Hill was right behind him. That's when Schumacher made an uncharacteristic error. He ran wide on a left-hand turn and his tires made contact with the wall. Hill saw an opening and darted for the inside line on the very next right-hander.

Only Schumacher did exactly the same. It was either a racing maneuver or pure cynicism to take out his rival, depending on who you ask. His Benetton plowed into the Williams, went momentarily airborne, and nosed straight into the tire wall.

Schumacher's race was clearly over, meaning that all Hill needed was a top-five finish to secure the title. But Hill's car had sustained damage too. He drove back to the pits dragging his left front wheel, only to be told that the wishbone, which connects the wheel mount to the car, couldn't be salvaged.

Hill never accused Schumacher of cheating. Then again, he didn't need to—the British press was all over it. *Bild*, Germany's most popular tabloid, responded to the barbs with a story on its front page about its hometown world champion.

"War of hate against Schumi," the paper wrote. "It is only envy."

Three years later, and driving for Ferrari, Schumacher was involved in another final-day crash with the championship on the line—surely a coincidence.

"He would chop someone. He would force his way through. He would show who's boss on a corner," Brawn says. "If you were racing Michael Schumacher, you knew you've got no quarter. If he saw a gap, if you left an inch, he would take a foot."

Once again, Schumacher came into the last Grand Prix with a one-point advantage in the standings. And once again, Schumacher was leading the race with a Williams on his tail. But this time, a bleach-blond French Canadian named Jacques Villeneuve was at the wheel. With twenty-two laps remaining, Villeneuve launched himself into a gap on the inside of a right-hand corner. Schumacher threw his wheel clockwise and the front right tire of his Ferrari went straight into the Williams.

Whatever he was trying to do, it didn't work. Schumacher spun off while Villeneuve sustained minimal damage and claimed the title. And this time, not even the German press could defend him. "No question about it—Schumi wanted to push Villeneuve out," *Bild* wrote.

These were, the *Frankfurter Allgemeine Zeitung* said, "Wild West manners."

That wasn't the end of it. Weeks later, the FIA took a closer look at Schumacher's actions and ruled that they had been deliberate. The organization stopped short of suspending him, preferring instead to strike him retroactively from the 1997 records and sentence him to a sort of community service. Michael Schumacher, of all people, would have to participate in a Europe-wide road safety campaign.

Schumacher's ethics were certainly questionable, yet no one could debate his commitment. At Benetton, the team realized it had no use for a test driver on a full-time contract since Schumacher insisted on driving the car as often as possible. He didn't let up once he moved to Ferrari either. In the winter before the 1996 season, Schumacher traveled to Portugal for his first test of the new car. The mechanics rolled up to the circuit at 8:15 a.m., as usual. When they arrived, Schumi was already there, in his race suit, sitting on the steps of the motor home.

"If you want to start winning," he told them, "you have to get up early in the morning."

Schumacher never seemed to get tired either. While other drivers staggered to the podium after a race, barely able to keep themselves upright, Schumi made a show of leaping about like Michael Jordan, explicitly to intimidate them with his physical prowess. And after that, he usually had enough energy left for karaoke.

No one in F1 history had ever taken fitness quite so seriously. Schumacher was the first to truly understand that being an F1 driver meant living as a professional athlete. He worked out two to three hours a day, every day, all so that he could be sharper in the car, where the driver was permanently engaged in a physical struggle to keep it on the road. Working the pedals amounted to sitting at a leg-press machine, and turning the steering wheel felt like maneuvering a manhole cover, all while losing up to four pounds in water weight. Schumacher's legendary training regimen included cardio, plenty of core work, and a specific focus on developing his neck to cope with the immense G-forces that come with driving a Formula 1 car. Since his debut as a skinny hothead, Schumacher had gone up three collar sizes.

There was also the offseason he spent in the Persian Gulf training in warm weather, while all of his rivals went on vacation. Or the time he boasted about hitting the gym with Sylvester Stallone. Whenever he could, Schumacher also played soccer for a local amateur team near his home in Switzerland. And anytime he boarded his private Learjet, he made sure that his mountain bike went along too.

Shortly after his move to the Scuderia, the Italian-based Technogym company equipped Schumacher with a half-million-dollar mobile gym built inside a truck. The dungeon on wheels followed him to Grands Prix and wherever he wanted to test. Ferrari was prepared to accommodate him. Whatever Michael wanted, Michael got, including Enzo's old office at Fiorano, equipped with a bed and a shower so that Schumacher didn't have to go home after a long day of practice.

There was only one accoutrement that Schumi felt was lacking after his first season in Italy, a single piece that Benetton had that Ferrari didn't. For 1997, Schumacher said he required Ross Brawn. Montezemolo and Todt duly obliged and went to get him. Back then, there was unlimited testing in Formula 1, so once Brawn arrived, the driver and technical director were able to pick up right where they'd left off.

As soon as they could, Schumacher and Brawn holed up at the test track for a week.

THE THING THAT MADE ROSS BRAWN SUCH A PRIZED COM-modity in the Formula 1 pit lane was the same thing that made him such a rarity among the sport's leading technicians. He wasn't a design genius who dreamed up wild ideas at a drawing board or an aerodynamics expert with a preternatural sense for airflow around a high-speed automobile. Rather, the soft-spoken Brit with a penchant for munching bananas on the pit wall had a specific gift for the management of a Formula 1 team.

At a time when they employed upwards of five hundred staffers in increasingly specialized roles, Brawn knew better than anyone how to organize their collective power to build a faster car. He seemed to possess an instinctive understanding of each individual's job. It turns out that was no coincidence—he'd actually done most of them.

Brawn began his F1 career as a machinist under Patrick Head at Williams back in 1978, as one of just eleven employees in the old carpet factory. Over the following two decades, as he climbed the corporate ladder, he would go on to hold virtually every position in the sport: mechanic, R&D technician, aerodynamicist, lead designer, racing strategist, and technical director. Brawn was the Formula 1 equivalent of the guy who worked his way up from the mailroom to the boardroom.

That experience had also given him an unparalleled knowledge of the arcane regulations and specifications that make up F1's rulebook. Which is precisely what Luca di Montezemolo was looking for.

"Our technicians need to interpret the rules in a more aggressive and 'extremist' way," the Italian would tell anyone who'd listen. "Less conservative."

If there was one thing that Brawn could never be accused of during his thirty-five years in Formula 1, it was being too conservative with the rules. A man responsible for some of the most creative—and contentious—innovations in F1 history, Brawn spent so much time hunting for loopholes and gray areas that the margins of the rulebook may as well have been his office.

The two world titles he delivered for Schumacher at Benetton had been accompanied by a steady drumbeat of complaints about supposed violations. Some of that was the classic bending of technicalities that every team engaged in. And some of it . . . went a little further than that.

Even before the first race of 1994, there were rumors that Benetton's car was powered by something more propulsive than a strapping young German with a thermonuclear competitive streak. Those suspicions were confirmed midway through the season when an FIA investigation discovered that the B194 car was equipped with a secret, hidden, and highly illegal launch control system.

The investigators found that the system, designed to optimize a car's getaway from a standing start, could only be activated via a concealed menu on the pit lane computer, followed by an elaborate series of up and down gear shifts by the driver in the car. This being 1994, it was no great mystery what was going on. Anyone who'd ever fired up their Super Nintendo and hit *up, up, down, down, left, right, left, right, B, A, Start* could tell you: Benetton's car literally came with its own cheat code.

Brawn disputed all of this, naturally. Sure, the car had a secret launch control feature. So what? It was only there for testing, obviously.

No one on the team had the faintest inkling that scrolling down past the final item on the menu screen to a hidden "Option 13" setting could activate the launch control right before a race. And as for that sequence of up and down gear shifts in the car? I mean, that was just to ensure that no one switched the thing on *by accident.*

Benetton somehow presented this defense to the FIA with a straight face.

Everyone involved knew that these excuses strained the limits of credulity. Benetton got away with it anyway, because Brawn had spotted an opening. The rules at the time only prevented the *use* of traction control, not the existence of software that *could be used* to enable traction control. And since the FIA couldn't prove definitively that Benetton hadn't just outfitted its car with a secret "Option 13," but actually deployed it in a race, well, there wasn't much the FIA could do about it. Benetton escaped with a $100,000 fine.

The expensive lesson Brawn took from the episode wasn't "don't do it again." In fact, it was the opposite. He came to understand that there were thousands of precise specifications and measurements in the F1 rulebook—but none of them concerned how far the envelope could be pushed. Even when they had you dead to rights, you could usually find enough wiggle room to weasel your way out of trouble. If it wasn't explicitly spelled out in black and white, then it was a gray area. And as Brawn looked at the sport's growing list of regulations, he saw a vast gray ocean of opportunity.

That realization was one trade secret he could take with him to Ferrari.

THREE YEARS AFTER MOVING TO MARANELLO, BRAWN AND his team had finally developed a car capable of winning a title. The only setback was losing their driver.

With a quarter of the 1999 season gone, Schumacher led the drivers' standings until a brake failure on the opening lap of the British Grand Prix flung him into the tire wall. It took nine minutes to extricate him from the shattered Ferrari. Schumacher suffered a double fracture below the right knee, effectively ruling him out of the remainder of the season and ending his title challenge.

It didn't end Ferrari's challenge, though. Because contrary to popular belief, Schumacher did have a teammate—an Irish playboy named Eddie Irvine who lived on a yacht. It's just that *teammate* didn't quite capture the full scope of Irvine's responsibilities. He was something closer to a subordinate.

That pecking order came from Brawn's rethinking of the standard driver setup that had existed since the 1950s. He wasn't interested in simply sticking two guys in identical cars and pitting them against each other. The way Brawn saw things, having two of your highest-paid employees at each other's throats just wasn't an optimal allocation of resources.

Rather than having two drivers competing for distinct individual goals, surely it was more efficient to have two drivers working together to achieve the same team goal. In this case, Ferrari's objective was clear: win Michael Schumacher a world championship.

So instead of being asked to race Schumacher, Irvine was asked to assist him, running his own car at a controlled pace to build a cushion to the rest of the field. In the pit lane, rivals teased him as Schumacher's butler. But Ferrari found a way to persuade Irvine that buttling for Schumacher wasn't so bad after all. In

1998, during negotiations over a new contract, Ferrari inserted an unprecedented clause: Irvine would be the first driver to receive a bonus if his teammate won the title.

With Schumacher missing for most of the season, Irvine took over Ferrari's campaign. And while he couldn't bring home a drivers' crown, the Scuderia settled for winning its first constructors' championship since 1983. That was enough to convince Ferrari that it was on the cusp of something big. The team just needed to figure out what could now push it over the top.

Ross Brawn, the guy who had held practically every job on the factory floor, had a knack for knowing exactly where to look for the next game-changing advantage. And this time, his instincts led him to what may have been the most head-smackingly obvious place imaginable—the only part of an F1 car that actually touches the road.

"What was apparent," Brawn wrote later, "was that the tire could be massively influential."

For most of Formula 1 history, the only influential episodes involving tires came when they spectacularly blew up. Even as designers and engineers radically reimagined every other piece of equipment in a bid to shave off an extra fraction of a second, the humble piece of black rubber that connects the car to the ground remained all but an afterthought.

That's mostly down to the fact that the tire is one of the few essential components that F1 teams don't engineer for themselves. They get them delivered from the same tire outfitters as everyone else.

For decades, Goodyear supplied a standard set of synthetic rubber tires to almost every team on the grid. There were different compounds to choose from, ranging from soft (quicker, less durable) to hard (slower, more durable), as well as two sets for damp or wet weather. The only material way that a team could affect the outcome of the race through tire choice was in deciding which of those to use at any given moment.

But in 1998, for the first time anyone could remember, Formula 1 had introduced major new rules around tires, mandating grooves in the front and rear tires as part of a broader effort to slow the cars down in the post-Senna safety drive. Goodyear, which saw little commercial uplift in being associated with tires deliberately designed to make your car go *slower*, promptly quit the sport, paving the way for the Japanese manufacturer Bridgestone to enter in 1998.

Like any other F1 newcomer, Bridgestone took some time to get up to speed. The company struggled to produce grooved tires that supplied the necessary grip and didn't fall apart within a few laps. Even after two seasons as the sport's official tire supplier, the verdict on Bridgestone from F1 drivers was "not so favorable," recalled Hirohide Hamashima, the firm's director of tire development.

So when French giant Michelin announced in late 1999 that it was returning to Formula 1 as a rival to Bridgestone, most of the teams couldn't wait to ditch the Japanese manufacturer and sign up for something closer to home.

Not for the last time, when everyone in F1 zigged one way, Ferrari zagged hard in the opposite direction. The team kept its chips on the other side of the world as the only title contender to stick with Bridgestone. This was the opportunity that Brawn had been searching for. Tires would become Ferrari's unfair advantage.

At Brawn's direction, Bridgestone operated less like the team's tire supplier and more like a brand-new division of Ferrari. An entire team of Japanese engineers moved to Italy, while Ferrari sent its own crew of engineers to be based full-time in Japan. Nothing was confidential or off-limits. "Everything you want, every piece of information that is in the company, you can have it," Brawn told the men from Tokyo. "And we want the same from you."

Ferrari armed Bridgestone with all of its chassis, engine, and suspension data. In return, Bridgestone agreed to hand over proprietary information on its compound and construction methods.

Never before had a Formula 1 team been so deeply involved in the production of its own tires. And as Brawn dug deeper into the process, he discovered that the tire was an engineering challenge unlike anything else in the sport. The goal for every other part of the car was to make it perform faster. The goal in building a tire was to make it fall apart slower.

Until now, Brawn had understood that challenge in terms of the central trade-off between grip and durability. More grip allowed you to go through corners faster, but wore down the tire more quickly. Less grip forced the driver to take the corner slightly slower, but meant the tire lasted longer. The Bridgestone engineers explained that it was a little more complicated than that.

A multitude of other factors played a role in tire degradation, from the setup of the car's suspension to the driving style of the guy sitting behind the wheel. And mathematical modeling could only get you so far. To really understand how a specific tire compound would perform over a seventy-lap race in different conditions—wet weather or dry, hot track or cool, at high speed or low—there was only one reliable method. You simply had to test it, over and over and over again.

Luckily, Ferrari had just the man for that. Schumacher tested the Bridgestone tires relentlessly. He logged so many laps on so many different rubber compounds, providing the Bridgestone engineers with so much invaluable data, that the Japanese company eventually agreed to pay for two additional Ferrari test cars.

What Bridgestone did with those terabytes of data was construct the most highly customized hunk of rubber compound the sport had ever seen. While Michelin produced its tires to satisfy half a dozen teams, Bridgestone provided Ferrari with a tire individually designed for the handling and suspension geometry of a Ferrari piloted by Michael Schumacher. It was like a bespoke suit from Savile Row, while the rest of the grid wore off-the-rack.

"You would spend weeks and months in the wind tunnel and get half a second per lap gain," Brawn writes. "But we could put on a new tire and get half a second just like that."

Brawn set about turning that half-second advantage into an unassailable dynasty.

The tire underpinned Ferrari's entire racing philosophy. Everything from the design of the chassis to the team's in-race strategy was conceived to leverage the characteristics of Bridgestone's high-grip, high-degradation tire. Ferrari developed a car with a smaller fuel tank and embraced a multi-pitstop strategy that turned the essence of Grand Prix racing on its head. Instead of a seventy-lap marathon, Ferrari turned it into a series of short sprints.

No one else could keep up. In 2000, Schumacher finally broke Ferrari's world drivers' title drought following a final-race show-down with McLaren's Mika Häkkinen. But once Bridgestone began to focus on Ferrari, Schumacher was uncatchable. In 2001, Schumacher retained his title with a record margin of 58 points. The next year, he smashed that mark, taking the championship by 68 points following a season in which he finished on the podium at every Grand Prix. Then in 2004, Schumi broke his own record for race wins, taking thirteen of eighteen races.

None of those victories offered a starker picture of Brawn's ingenuity than the 2004 French Grand Prix at Magny-Cours, one of the rare occasions that season when Schumacher found him-self trailing a quicker rival—a fresh-faced young Spaniard named Fernando Alonso.

Ferrari pivoted on the fly to a radical strategy that called for Schumacher to make four separate pit stops, normally a sign that something has gone horribly wrong. Schumacher's F2004 was functioning perfectly, of course. But Brawn and his team calcu-lated that by keeping Schumacher on practically new tires, the team could both make up for any time lost in the pits and negate the slight pace advantage of Alonso's car.

By the time Alonso and his team cottoned on, it was already too late. Ferrari had turned the seventy-lap race into five flat-out sprints, each between eleven and eighteen laps. Despite spending nearly a minute and a half with his car jacked up in the pit lane, Schumacher won by eight seconds.

That season, Schumacher cruised to his fifth consecutive title and seventh overall. And after clinching the championship in Japan, near the home of Bridgestone, he could finally let his hair down. He stole a forklift truck and threw a fridge through a window. Unfortunately for Schumacher, an enterprising pho-tographer captured the whole thing on film and images of the incident were published in *The Sun* under a headline that read: "Schu Trouble Macher."

Somehow even that turned into a victory for Michael. Rather than tarnishing his image, readers said it made him more relatable and more human. For rivals, the Schumacher and Ferrari era had been completely dispiriting. But for the sport of Formula 1, which knew that a Ferrari revival would be good for business, it was a dream come true.

Until it wasn't.

As Schumacher's wins piled up and the margins of victory grew wider, Ferrari ceased to be a feel-good comeback story and began to look more like Formula 1's endgame. The Scuderia's budget of $400 million was already double that of McLaren— and four times what Williams was spending. Ferrari was the most popular team, the richest team, and now it was also the most successful.

This was the central paradox of a successful Scuderia. Everyone wanted to see the Prancing Horse taking checkered flags. But once it became the only thing they saw, they realized they were no longer watching races at all. Attendance fell by as much as 50 percent at some venues, TV ratings slumped, and sponsors began to question their commitments. One British newspaper suggested the *Cavallino* logo should be replaced by a python, since Ferrari was slowly squeezing the life out of Formula 1.

"For five years, we won every race," Brawn said. "It was predictable."

In Bernie Ecclestone's mind, predictable was just another way of saying boring, motor racing's cardinal sin. And when Formula 1 gets boring, Ecclestone says, there's only one thing you can do: "Find a nice church to go and pray."

Failing that, you could find Max Mosley and have a chat, which is exactly what Bernie did. Ahead of the 2005 season, the FIA announced out of nowhere that tire changes would be banned going forward. At a stroke, Ferrari's advantage—all of that work with Bridgestone, all those laps testing rubber compounds, and all the multistop strategies—had gone the way of

Williams's supercomputers. The FIA called it a safety measure. Brawn could tell what it was really about.

"It was an odd rule," he said, "but just done purely to screw Ferrari."

THE ODD RULE SUCCEEDED. FERRARI WAS SCREWED.

From the opening races of the 2005 season, it was clear that the Scuderia's reign was at an end. The team looked slower and struggled with reliability, while a rejuvenated Renault sprinted away from the field. Ecclestone and Mosley had their wish.

What they hadn't foreseen was that messing with tire rules would have a profound impact on the nature of the sport. Not only had their clumsy maneuver stopped F1's most popular (if painfully dull) team in its tracks. It also teed up the most humiliating show Formula 1 had ever organized. That July, on a hot weekend in Indianapolis, 120,000 people bought tickets to watch the world's premier motor racing series. The absurd spectacle they got instead was a two-hour parade of just six cars.

"That was a tough moment in Formula 1," says Stefano Domenicali, who was just one more embarrassed Ferrari director in Indy, years before becoming CEO of Formula 1. "That's really what we can't forget."

The self-inflicted debacle had begun on an otherwise quiet Friday afternoon during free practice. Ralf Schumacher, Michael's kid brother racing for Williams, was motoring through Turns 12 and 13 of the circuit when he went hurtling into a wall at 190 mph. The accident would rule him out of the race on Sunday.

But this wasn't any ordinary crash. The issue was a catastrophic failure of his left rear tire that looked eerily similar to another driver's tire problem earlier in the day. Whispers along the pit wall suggested they weren't the only ones. By Friday evening, mechanics up and down the paddock found that they were seeing the same problem on at least half a dozen other cars: long vertical cuts in the sidewalls of their tires. And those cars had one thing in common. They were all running on Michelins.

The three teams on Bridgestone rubber—Ferrari, Jordan, and Minardi—saw no apparent issue. But for the seven outfits on Michelin, this was blistering into a full-blown crisis. Back at headquarters in Clermont-Ferrand, France, chemical engineers began a furious battery of tests to work out what was going wrong. As it turned out, the French tires were uniquely ill-suited to the banked sections at America's most famous circuit. On Saturday, Michelin made the terrible admission that its product was unsafe for the Indianapolis Grand Prix.

In a panic, the company suggested flying in a new batch of tires for the race. But the rulemakers told them this would be unacceptable. The seven Michelin teams came back with a counterproposal. What if the circuit added a chicane around Turns 12 and 13 to force drivers to lift off the gas pedal? The lower speeds would put less stress on the tires and keep everyone from blowing up. As crisis talks progressed, nine of the ten teams found this to be a reasonable idea. The one holdout was Ferrari, brought to you by Bridgestone. And as usual, Ferrari and the FIA seemed to be on the same side.

"To change the course in order to help some of the teams with a performance problem caused by their failure to bring suitable equipment to the race would be a breach of the rules and grossly unfair to those teams which have come to Indianapolis with the correct tires," the FIA explained.

On Sunday morning, the matter still hung over the Brickyard like the stench of exhaust fumes and light beer. Officials reminded the Michelin teams that they had three options. They could switch tires and incur a penalty. They could make a bunch of time-consuming pit stops and accept further penalties there. Or they could simply drive slower through the problematic corners.

Bottom line: there wasn't really a choice at all. In a fractious meeting with Bernie Ecclestone, the furious teams threatened not to participate. Bernie warned them that if they pulled out, then that was up to them. They just needed to realize that 120,000 people would be directed to them for refunds. "I don't know how you guys are all going to get out of here if you don't race," he said to the teams. "So my advice to you is you should start the race . . . But if you decided you wanted to stop after two or three laps, there's nothing I could do to stop you."

In his own backhanded way, Bernie had given them an out. The seven Michelin teams didn't think twice about taking it. Though all twenty drivers started the formation lap, only six lined up on the grid for the race start. The other fourteen, fitted with Michelin tires, all drove back into the pits in protest. For the first time in more than fifty years of racing, a Grand Prix would go off with fewer than ten entrants.

Under a hail of boos, bottles, and jeers, an untroubled Michael Schumacher drove the seventy-three laps to victory. Only a fraction of the crowd stuck around to see him take the checkered flag. "It was not the way I wanted to win my first one this year," Schumacher lied.

That evening, while attendees queued at the box office for refunds and police tried to quell the angry mob, Ferrari flew home with the first-place, second-place, and constructors' trophies. It didn't feel good. Going with Bridgestone had brought the team success once again, but these would never be the most prominently displayed bits of silverware in Maranello. Formula 1's reputation, at least in America, had suffered irreparable damage.

"We understood that for the bigger picture it was a problem," Domenicali says. "The fans felt betrayed."

7

American Exceptionalism

THE FIASCO AT INDY WAS the last straw.

Though the US Grand Prix returned to Indianapolis the following year, in 2006, Bernie Ecclestone's mind was made up. He had tried for years to tap the sport's largest open market, knowing that F1 needed to keep changing in order to grow, but his European road show simply didn't mix with America. The presence of other major motor racing series was already a drag, interest from television was cool at best, and the culture clash of selling old-word luxury in places like the Brickyard was proving too much to overcome. A dispute over tire safety was merely incidental.

Enzo Ferrari may have recommended treating Americans like hicks, but Bernie realized there was another way: he could avoid dealing with them altogether, a position he would reiterate to anyone who'd listen in years to come. "I'm not very enthusiastic about America," he once told Russian television. "The biggest problem with America is they believe they are the greatest power in the world. Not in reality, but in belief."

Ecclestone had felt that way from the moment his awful trip to Indiana began to unravel. "The future of Formula 1 and Michelin in the United States is not good," he said at the time.

The truth was that the past of Formula 1 in the United States wasn't so good either. The whole adventure had been one false start after another, ever since the sport first visited America in the 1950s. Eager to showcase their European know-how in one of the world's largest and most profitable luxury car markets, teams crossed the Atlantic to the United States for an F1 Grand Prix at Sebring International Speedway in Florida in 1959. It didn't exactly turn into an ad for reliability: only seven of the nineteen entrants completed the race.

The following year, organizers took them even farther afield for another forgettable event in Riverside, California, before the US Grand Prix began a two-decade residency at Watkins Glen in upstate New York. The race still wasn't making much money—in fact, the teams were flat-out losing money by competing there in the 1970s—but at least it had a home. The problem was that broadcast fees were hard to come by, and unlike in Europe, the race couldn't count on any government subsidies. The bulk of the Grand Prix's income relied on ticket sales, all for a race that was 250 miles northwest of New York City and closer to Canada than the bright lights of Times Square.

So it didn't surprise anyone in 1980 when the company behind the crumbling circuit simply ran out of money. Ecclestone, who had no patience for anyone who couldn't pay him, was all too glad to yank the Grand Prix away and try somewhere with a little more pizzazz than upstate New York.

That somewhere turned out to be Las Vegas. The city founded by ranchers and transformed by the Mafia seemed like a much more natural landing spot for Ecclestone. Vegas was still in its relative infancy as a major destination, but the early monuments of Sin City were already in place: the Flamingo, the Sahara, the Tropicana, and that hulking edifice of in-your-face kitsch on the Strip known as Caesars Palace. That's where Bernie set his sights.

In the hopes of taking a Grand Prix to Nevada in late 1981, the F1 executives had already buttered up the reps from Caesars at a lavish London party inside the Royal Albert Hall. Later, when Ecclestone flew out to the States with Max Mosley to finalize the arrangements, Caesars returned the favor, putting them up in some of the hotel's largest and most garish suites. Only after

Mosley walked in did he realize that his room came with a giant mirror on the ceiling.

Looking up, he naïvely rang Ecclestone's room to ask what it was for. Bernie didn't miss a beat in spelling out precisely what this town was all about.

"It's so you can comb your hair in bed," Ecclestone told him.

Caesars Palace truly did have everything—at any hour and for any taste. But what most interested Bernie was the sprawling parking lot, which seemed to be larger than some European countries. For one weekend in October 1981, Ecclestone imagined it as a Formula 1 circuit. The end product was a track that squiggled through the lot, out into the desert and back again, with a few insanely high-speed corners, and an ambient temperature that slow-broiled drivers in their cars. Bernie couldn't persuade many fans to come out and he still couldn't gin up any interest from TV. The weekend was such a bust that when Nelson Piquet brought his Brabham home in fourth place to clinch the title, Ecclestone wasn't even interested in celebrating with his own world champion driver.

Instead, Bernie hit the tables at Caesars and landed in a huge card game with a group of elderly Chinese women. He promptly lost $100,000. America and Ecclestone just weren't getting along.

"The Americans wanted it to be American," he says, "not the way it was."

STILL, FORMULA 1 KEPT GOING BACK, SEARCHING FOR THE right conditions to make something stick in the one major territory it had yet to conquer. In 1982, the United States became the only country on the calendar to host three Grands Prix—on the streets of Detroit and Long Beach, California, and once again in the parking lot of Caesars Palace. There were even plans for a New York Grand Prix in 1983, to be held either at the Meadowlands in New Jersey (home of the NFL's Jets and Giants), in Queens near LaGuardia Airport, or on Long Island with the city skyline for a backdrop. But the plans fell through in a matter of months. By

1985, after F1's brief and chaotic flirtation with Dallas, Detroit was the only US race left in the lineup. Like every other American venue, though, it was soon replaced.

From 1989 to 1991 it was Phoenix, Arizona's turn as the eighth different US city to try to make a race work. It only succeeded in becoming perhaps the most forgettable Grand Prix venue of all time. The pit lane was a mess. Barely fifteen thousand fans showed up. And in 1990, the world's premier motorsport was outdrawn by a rival event on the outskirts of town: the Chandler Ostrich Festival, where seventy-five thousand people turned out to watch crazy-looking flightless birds race each other instead of F1 cars.

The moment Ecclestone could tear up Phoenix's five-year contract, he did. "I tried to save the race," he said at the time. "The people of Phoenix just didn't put in the effort. I don't know why they bothered in the first place."

On to the next disaster, then—one where Formula 1 hopefully wouldn't have to compete with an ostrich jamboree. Ecclestone kept scouring his list of American promoters. For such a big country, there weren't really a lot of suitable options. Besides, he'd already pissed off most of the decent ones.

"We need a race in the USA, but it has to be in the right place," Ecclestone explained in 1992. "Our next marriage has to be forever."

Bernie, who was personally on his second of three marriages at the time, did genuinely think that Indianapolis could be the one. A few trips to the Brickyard, however, soon showed him that his affections were misplaced. Indy had held its first Grand Prix in 2000, but by 2005 the groundswell of interest still hadn't materialized the way Ecclestone hoped.

In particular, the American reluctance to put F1 racing on network TV had turned him off. One of his stipulations as far back as the European Broadcasting Union deal in the 1980s was that the sport had to be on free-to-air stations. But now he found that no broadcaster in the United States was willing to bite. "The Americans really only wanted to watch the Monaco race," he said. One creative solution was Ecclestone's partnership with the American superagent Casey Wasserman to package four Grands Prix and buy airtime for them on CBS—as if they were placing a giant advertising spot. But even that turned out to be half-baked. The last of the four races wasn't even shown live. The few fans who bothered to tune in a month after the fiasco at Indy only found the German Grand Prix on tape delay.

Promoters were frustrated too. The Indianapolis Motor Speedway was shelling out $15 million in fees to host the Grand Prix and wasn't even recouping it in ticket sales. The blame, Ecclestone argued, lay solely with Indy's organizers. "You can move into town and you can't tell there's an event going on, and that's the problem," Ecclestone said. "It should be promoted aggressively, and they've understood that at all the places where Formula 1 has been successful. They don't seem to get behind that here."

Ecclestone had no idea how much worse 2005 would make it. The sport had blown its latest chance to impress America. And America wasn't sad to see it go.

"F1 is the rude houseguest who never brings anything to the party and continues to wipe its muddy shoes on the new Persian rug," the *Indianapolis Star* wrote on that Monday morning in 2005. "And to do it at Indianapolis Motor Speedway, the mecca of all motor sports . . . They had no respect for the place, and, clearly, no respect for the fans.

"Goodbye, Formula 1," it added. "Good riddance, Bernie Ecclestone."

THE SECRET TO AN ECCLESTONE F1 CALENDAR WAS NEVER LET-ting anyone know precisely where they stood. Sure, there were contracts and theoretical dates, but like everything else in the sport, whole chunks of the season could change from one year to the next. The Bernie method was to make sure that no race was ever safe.

Technically, the schedule was always published by the FIA, the final decision maker over the rules and integrity of the sport.

In practice, though, Formula 1 raced wherever Bernie told it to race, which in 2005, meant a nineteen-race slog spread over seventeen countries. Through endless talks with local promoters and circuits, he made sure that there were always more interested hosts than available slots. That way, he always had a fallback. If Rio de Janeiro didn't play ball, then Bernie had São Paulo at the ready. Over his decades in charge, the calendar swelled by more than 50 percent.

The genius of it was that there was no such thing as a flat rate for hosting a Grand Prix. Ecclestone could look at the market, decide how essential it was to promoting F1 (or vice versa), and name his price. Even as far back as 1973, when Bernie was merely the envoy of the British constructors, he was convincing European circuits to put up around $56,000 for hosting rights and the non-European circuits $110,000. Soon, those numbers would grow to eight figures. And no one's money was too tainted to take.

Long before the leaders of global soccer or the Olympic movement routinely sold their souls to the highest-bidding regime in the 2010s and 2020s, Formula 1 was selling Grand Prix rights to anyone with enough cash and the space for a circuit. No one did much hand-wringing about the role sports might play in varnishing the images of dictators. The truth was that Ecclestone and the FIA didn't mind visiting an authoritarian host country—in fact, they quite liked it. Those leaders didn't have to worry about popular opinion at home, or marshaling public funds to make projects happen. They just got things done.

So without a second thought, Ecclestone and the FIA spent years acquiring a rogues' gallery of global partners. In the 1970s and '80s alone, Bernie's Formula 1 raced in Juan Perón's junta-controlled Argentina, apartheid South Africa, and Communist Hungary, where the event officials reported directly to the KGB.

"The only thing I knew was they were more honest people and easier to deal with," Ecclestone says of his Kremlin-connected counterparts. "They knew that I'm proud to say I'm a handshake guy. I don't worry too much about contracts. I sign things anyway without reading them."

Unlike other sports in years to come, the FIA would never pretend that they were taking the sport to these places for humanitarian reasons. They weren't there to "shine a light on injustice" or "bring the world together." Any talk of the local regime was strictly taboo, even if it was condemned by the United Nations, as was the case when F1 returned to the Kyalami circuit north of Johannesburg in 1992.

"We have eighty countries represented in the FIA; we have forty-five million members; we do not ask that we should like them all," FIA president Jean-Marie Balestre had said during a previous visit to South Africa. "None of them want to find themselves politicized like the Olympic Games."

For every dictatorship or human-rights-abusing government visited by Formula 1, however, there were plenty more hosts with a more damning flaw for the sport: they never wanted to pay enough. This was mostly the case for the collection of tracks dotted around Europe, the heartland of open-wheel motor racing, where circuits hoped that decades of tradition might earn them a discount.

Monza, for instance, had hosted motor racing events since the early 1920s, back before Ferrari and Mercedes even existed. Save for a year of renovations in 1980, the Autodromo Nazionale has been in every single season of F1. It is the de facto home race for Ferrari, and if it were to be dropped from the schedule, the Italian government might start a war. For that, Formula 1 was willing to be a bit more flexible.

This is the eternal balance the sport has to strike in selecting circuits. There are heritage tracks and there are moneymakers. During the Bernie years, F1 visited more than forty different venues—and that was only a fraction of the events that materialized. Ecclestone was willing to push the boundaries even further. In the 1980s, he put together a proposal for a Moscow or St. Petersburg Grand Prix for Soviet premier Leonid Brezhnev. Then, twenty years after that went nowhere, he negotiated with the city of Moscow again. At which point even the Russians found Bernie a little unsympathetic to deal with.

"Ecclestone wanted to keep all the rights on it—the rights to ticket sales, to television broadcasts, to advertising," the mayor of Moscow said at the time. "If we had agreed to his terms, we would have got nothing but the exhaust fumes."

There was only one place where Ecclestone found himself at a disadvantage year after year. It was the one Grand Prix he couldn't do without, because it was simply too wrapped up in everything the sport wanted to be. This was the race with the glitz, the glamour, the movie stars, and a literal prince in charge of it. Located on the French Riviera, the place that Somerset Maugham called "a sunny place for shady people," the Principality of Monaco boasted the one thing no other circuit could.

It had leverage over Bernie Ecclestone.

THE ENTIRE PRINCIPALITY OF MONACO, AN INDEPENDENT state ruled by the same family since Francesco Grimaldi claimed a rock on the Mediterranean in the thirteenth century, is about two-thirds the size of Central Park. That anyone ever thought it would be a good idea to race cars there is, frankly, absurd.

Mechanics try to give their drivers a little help by hiking up the seats in the cockpit by 10 or 15 millimeters, just to improve visibility, but that only goes so far. The cars tend to be more temperamental than usual around Monaco due to lower speeds that cause the engines to overheat. In other words, it's a circuit where everything can go wrong—and it usually does. Even Michael Schumacher, who won there five times between 1994 and 2001, had to recognize the madness of it. The German told it straight: Monaco was downright unsafe.

"For so many years we have successfully campaigned for more track safety and then we race in Monaco," he said in 2012, pointing out one of a gazillion contradictions in the sport. "But in my view this is justifiable once a year—especially as the circuit is so much fun to drive."

Really, it's only fun when you're in front, which Schumacher usually was. But qualify anywhere other than pole position on

Saturday, and your chances of crossing the line first on Sunday are slimmer than anywhere else. It's no coincidence that of the thirty race winners from 1984 and 2014, sixteen started in first. The challenge only became more difficult the bigger the cars got. The Lotus 18 that Stirling Moss drove to victory around Monte Carlo in 1960 wasn't even three-quarters the width of Ayrton Senna's winning McLaren MP4/5 in 1989. The years went by and the cars evolved, but the streets of Monaco stayed the same size. In many places, it became all but impossible for two cars to drive side by side. Crashes are inevitable.

That's why, despite being the lowest-speed Grand Prix of the year, no race on the schedule sees fewer cars reach the finish line than Monaco. In 1996, that number hit an all-time low with three as rain, crashes, and reliability issues conspired to create one of the strangest races in history. Every driver who completed the race also wound up on the podium, led by the out-of-nowhere Frenchman Olivier Panis. He had never won a Grand Prix before, nor did he ever win one again. The result was so unexpected that he hadn't even packed a suit for that evening's gala dinner with the prince—and this being a Sunday in Monte Carlo, Panis had nowhere to buy one. Luckily, one of the most powerful men in Monaco was on hand to help him out with a spare: he was the concierge at the Beach Plaza Hotel.

Yet for all of its drawbacks as a competitive spectacle, the Monaco Grand Prix established itself early on as an iconic selling point for all of Formula 1. And for that, the sport had a pair of 1950s newlyweds to thank. Their names were Rainier III, the 12th sovereign Prince of Monaco, and Grace Kelly, the American movie star. They married in a fairytale ceremony in 1956 and proceeded to turn their cliffside on the Riviera into one of the most glamorous corners of the planet. Rainier empowered the Automobile Club of Monaco (ACM) to organize and promote the Grands Prix, all to the glory of his tiny, wealthy tax haven of a country. And Grace brought Hollywood.

It didn't take long for her address book to double as the Grand Prix's official guest list. Nor did it hurt that the timing of the

Grand Prix broadly coincided with the Cannes Film Festival just down the coast. Everyone who was anyone already happened to be in the neighborhood, hoping for an invitation from the Palace to the unofficial opening of high season on the Riviera. By the 1960s, heads of state, studio moguls, actors, singers, hangers-on, and assorted international gangsters all knew that Monte Carlo was the place they had to be in late May. All they needed was a yacht, or, better yet, a suite at the Hôtel de Paris.

Monaco around race weekend would become the world capital of calling in favors. Who had a spare pit lane pass going? Was there a hotel room around—with a view of the marina, please? What about a place to dock in the port?

Frank Sinatra was such a close friend of Kelly's, and such a regular visitor to Le Rocher, that he became godfather to her daughter, Princess Stephanie. The Beatles visited the race in 1966, post–*Hard Day's Night* and pre–*Sgt. Pepper's*. The Rolling Stones came back for decades. And, for a little while, the Greek shipping magnate Aristotle Onassis was a fixture at the race aboard his yacht the *Olympic Winner* as one of the early investors in monégasque property—at least until falling out with Prince Rainier over Onassis's plans to turn the principality into Vegas-on-the-Med. Rainier, a shy man who much preferred to build his dominion's glitz around fast cars and blue water than gambling and green baize, wanted the Grand Prix to be Monaco's signature.

On that front, Prince Rainier and Princess Grace succeeded.

Generations of world champions moving to the principality to enjoy its clean streets and nonexistent income tax were all the endorsement they needed. In fact, Monaco is the closest thing to a home race for most drivers. Today, Lewis Hamilton is practically neighbors with Monaco native Charles Leclerc and their great rival Max Verstappen. They all get to sleep in their own beds for precisely one Grand Prix a year. The convenience is undefeated. When the heat shield on Kimi Räikkönen's car caught fire in 2006, he simply walked away from the smoldering carcass of his McLaren to the marina instead of the pits. Still wearing his helmet and firesuit, he strode up the gangplank of his yacht, the *No Name,* and spent the rest of the race knocking back beers.

For those who don't drive Formula 1 cars and need to pay for the privilege of attending the race and boozing on yachts, the spending spirals quickly. That's what happens when demand turns the principality from a thirty-thousand-person village into a two-hundred-thousand-reveler stadium for four days. But as the entire country's hospitality industry knows, this is not a crowd that tends to look at prices. Marlboro was spending a couple million dollars on entertainment in a single weekend as far back as the 1980s. More recently, hotel rooms are known to fetch north of $10,000 for just two nights. Occupancy hovers around 100 percent.

The Société des Bains de Mer, the state-founded, publicly listed company that owns hotels, casinos, and other property around the principality, claims that the Grand Prix weekend alone drives 5 percent of its annual profits. The reputational value of having F1 as the nation's postcard, meanwhile, is priceless.

To fans and drivers, all of that entangled, intangible history was what made Monaco special. To Formula 1, it's what made it essential. For more than sixty years, the Automobile Club and the principality used that to cut a deal with the sport's organizers like none other in motor racing. The biggest concession it extracted from F1 was editorial control over the television pictures, even throughout the Bernie era. The world saw precisely the version of Monaco that Monaco wanted it to see. That extended to the trophy ceremony at the end of the race, where the podium is always on the steps of the Royal Box with the monarch in attendance. Unlike every other Grand Prix, there is only one visible logo during the presentation and it doesn't belong to any sponsor—not an oil company or a beer manufacturer or a crypto scheme. Only the seal of the Automobile Club is permitted.

Everywhere else around the circuit, the signage and sponsor logos look more familiar. The Monaco deal, however, ensured that the ACM shared in the profits of those with Ecclestone's Formula One Management, a privilege no other Grand Prix promoter would dream of asking for. The advantages were so substantial that many people around the sport believed that F1 had waived Monaco's hosting fee altogether. Even in the fantasyland of Monte Carlo, that much was a myth. In reality, Monaco pays a historically low fee, believed to be around $15 million a year, on par only with Monza. Newer additions to the calendar can expect to pay three times that price.

Ecclestone would have loved to charge the principality more. Yet each time the contract came up for renewal, he knew there was only so far he could push them. He could threaten to cut Silverstone, the home of British racing, and he could joke about dropping a place like Hockenheim in Germany (which he eventually did), but Monaco and Monza were the closest things in F1 to untouchable. "You know, we'd race here for free," Bernie told Prince Rainier in one of their many private meetings. "It's worth so much for Formula 1. So really, in theory, we would race here for free. But don't try me."

That didn't stop Ecclestone from posturing in public. He was doing it as much to protect the prices that he was charging every other circuit as for his own reputation as a tough-as-nails negotiator.

"The Europeans are going to have to pay more money or we will have to go somewhere else," he told *The Independent* in 2010. "We can do without Monaco. They don't pay enough."

It won't shock you to learn that those comments were made at the height of renewal talks with the Automobile Club. Weeks later, Ecclestone signed the circuit to a ten-year extension. He understood better than anyone that he could make up the shortfall elsewhere.

In the early 2000s, he'd managed to add China, Malaysia, and Turkey (never mind the autocrats). But there was another frontier emerging. And those investors were about to make everyone else look basically penniless. Ecclestone wanted to take his circus into the Gulf.

TO UNDERSTAND JUST HOW SIGNIFICANT IT WAS WHEN FOR-mula 1 decided to venture into the Arab world for the first time, it's important to remember that in the age of oil money in sports, this was still predawn. Qatar, which hadn't even considered bidding for a soccer World Cup yet, was mostly known as an American air base. Abu Dhabi was years away from moving to buy Manchester City. And Saudi Arabia, rich as it was, remained one of the most shuttered societies on earth (its brief investment in the Williams team notwithstanding).

But in the early 2000s, one tiny Gulf country with a car-mad ruling family had already clocked that F1 could help put it on the map. Facing competition from Dubai, Egypt, and Lebanon, a collection of islands that sits opposite Iran elbowed its way to the front. Bahrain, a country that prided itself on exporting the best Arabian horses, was about to import some serious horsepower.

The surprising part, in retrospect, is that despite its oil-and-gas-driven economy, Bahrain wasn't yet a Middle Eastern outpost for the super-rich. In 1932, it had been the first place in the Gulf where American companies struck oil, years before black gold started flowing out of Saudi or Kuwait. But in 2000, Bahrain's GDP was only half that of Qatar's and ranked behind those of Serbia and Jamaica.

Instead, the country of barely a million people existed primarily to serve the interests of the House of Khalifa, which had ruled the archipelago since the 1700s—even when it was technically under British control after World War II. The fact was that Bahrain quite appreciated the presence of the British, who proved useful for propping up whoever happened to be in power. So imagine the Khalifas' disappointment when Prime Minister Harold Wilson announced that he would withdraw all UK troops "east of Suez" by 1971.

"Britain could do with another Winston Churchill," the emir of Bahrain said, after a group of sheikhs offered to pay the troops to remain in the region. "Britain is weak now where she was once so strong. You know we, and everybody in the Gulf, would have welcomed her staying."

In that vacuum of the 1970s, seven kingdoms banded together to form the United Arab Emirates. The original plan was for nine members—the two that pulled out due to infighting between the sheikhs were Qatar and Bahrain. They preferred to form their own independent states. But while Qatar bet its future on natural gas and the UAE successfully converted its economic offering from pearl diving to oil and real estate, Bahrain failed to grab any of the global spotlight.

It took a meeting between Bernie Ecclestone and a cousin of the crown prince, Sheikh Jaber bin Ali al-Khalifa, to change that. Bernie wanted a race in the Arab world and Sheikh Jaber wanted to promote tourism to Bahrain. He also happened to love fast cars and European engineering, a taste he'd developed when he trained as an F-16 pilot in the Royal Air Force. After that, the sheikh liked to joke that he'd once traded his camel for a Ferrari.

Construction on the Sakhir circuit, south of the capital city of Manama, began in late 2002 with a view to hosting a Grand Prix early in the 2004 season. As Bahrain's shop window to the world, everything would need to be first-class. There would be no crumbling garages like at Silverstone, or pockmarked access roads like Monza's—only the finest materials for the brand-new $150 million track. They sent a fifty-person delegation to study the 2003 Spanish Grand Prix. They erected permanent buildings on the paddock to save teams shipping their motor homes to the Gulf, and imported the grippiest track surface materials from a quarry in Shropshire. They even sprayed an adhesive substance all over the sand surrounding the circuit to prevent it from blowing onto the road. Questions of Bahrain's dreadful human rights record and protests in the streets of the capital were dealt with as one more inconvenience to be handled for F1. "Even though we had all these problems with people saying, 'They torture us,' and all this rubbish," Ecclestone recalls, "I said, 'Let me talk to these bloody people and try and explain how it's good for the country.'"

That was merely the beginning. Just as other sporting projects eventually would all over the Gulf, Formula 1 served as the perfect excuse for Bahrain to build state capacity. It extended the

road network with sparkling new highways and took a sixteen-
ton shipment from Siemens containing only new traffic lights and
the cables to connect them. A race for fewer than two dozen cars
was now reshaping the entire country's road system.

Races in Europe, meanwhile, were struggling to stay afloat.
While Sakhir grew from a mirage in the desert into one of the
finest racetracks anywhere in the world, Max Mosley was, in
2003, cutting mainstays like Belgium and Austria from the F1
calendar. Montreal soon went the same way. Not only was Bahrain a slicker facility, but the new circuit was also immune to
new regulations in Europe and North America that limited tobacco advertising. So when Schumacher won the inaugural race
there, ahead of his Ferrari teammate Rubens Barrichello, and
BAR-Honda's Jenson Button, the names across their chests on
the podium read: Marlboro–Marlboro–Lucky Strike.

More important, Bahrain showed the rest of the Gulf what
was possible. Under the original deal with Formula 1, the organization could only add one more Grand Prix in the region,
which turned out to be Abu Dhabi. The crown prince first sat
down with Ecclestone in early 2006, and Bernie told him why
he'd never be able to stage a race. "Your streets are all at right
angles," he said. "What you need is a really nice open space to
build a proper circuit."

By 2009, the emirate had found one. Abu Dhabi unveiled
a $1 billion track as the centerpiece to a $40 billion redevelop-
ment of a place called Yas Island in a lavish display of Ecclestone's
favorite quality: getting things done. It sat beside a crystal-blue
marina inspired by Monte Carlo and came adjacent to a sprawling covered theme park called Ferrari World where visitors could
"Find That Ferrari Feeling" by riding the Formula Rossa roller
coaster or park their kids at Khalil's Car Wash.

Clearly, Abu Dhabi viewed a weekend of Formula 1 as the
pointy end of a global branding strategy designed to reinvent the
country's image abroad. No longer would these be isolated, auto-
cratic corners of the desert with oil money and dubious human
rights records. They were stops on the most prestigious racing

calendar in the world. As long as the price was right, Ecclestone was happy to sell them that privilege, just as he'd always done.

There was only one difference in how Bernie viewed his job in those days—something that hadn't really come up in his first three decades of running F1. By the mid-2000s, richer than he ever imagined and into his seventies, he felt that the business of hawking slots on the calendar and media rights was becoming, honestly, a pain in the ass. Bernie was ready to cash in his biggest chips and sell off the entire sport.

8

Fire Sale

ON THE LONDON AFTERNOON IN 2005 when Bernie Ecclestone sat down to lunch with Donald Mackenzie, the cofounder of a private equity firm named CVC Capital Partners, there was only one thing on his mind—and it wasn't his usual concern over who might pick up the tab.

Bernie was there to tell Mackenzie all about the business of Formula 1. Specifically, he wanted to say, it was a bloody disaster.

The entire sport was in crisis, Ecclestone explained. The teams were always clamoring for more money. Credit was drying up. America wasn't panning out. And now the European Union was threatening to pull the rug out from under the whole thing by outlawing tobacco sponsorship. For the first time in thirty years, Bernie didn't have an ace up his sleeve either. Much worse, he had a boss on his back. Through none of his own doing, Ecclestone's largest partner in the racing enterprise had become one of the most irritating entities he could think of: a German bank.

Mackenzie listened carefully. The trained accountant could tell that the sport's ownership and governance were messy. All those Ecclestone handshakes and obfuscations had made it that way on purpose. But Mackenzie could also tell there was latent global popularity in F1. And where there was popularity, there was opportunity.

In nearly two decades of experience, CVC had invested in an eclectic range of industries, from international media to suitcase manufacturers. Its record wasn't always stellar. The company's reputation was for making big bets that worked out beautifully or failed spectacularly. Its compensation model only raised the stakes: it rewarded executives with huge payouts if deals went well but forced them to cover certain losses personally. In 2005, CVC lost 240 million euros on a concrete materials manufacturer and face-planted on a stake worth over $1 billion in an Australian broadcaster.

But Mackenzie knew a distressed, undervalued asset when he saw one, and Formula 1 was just that. Over in the Premier League, clubs that had been on the brink of bankruptcy were starting to sell for nine figures. Pay television, which underwrote English soccer, was going from strength to strength. And foreign investors were beginning to sniff around many of the UK's quintessential sports products. In Formula 1, CVC saw a business with a bull of an engine that was limping along like its tires were slowly going flat.

So Mackenzie gave Ecclestone a soft pitch. He told Bernie to ring a man named Carmelo Ezpeleta, who was the Bernie of Moto GP, the top motorcycle racing series in the world. CVC had been the majority shareholder in Moto GP since 1998 and Ezpeleta could give him an idea of how the fund liked to operate. They'd straighten out the business, add some value, and stay out of Bernie's white hair.

"If you like what you hear," Mackenzie told him, "let's see what we can do."

HOW THE SELF-MADE FORMER CAR DEALER TURNED RACING promoter came to sell his pride and joy to a private equity firm—while attempting to stay in charge anyway—is a story that had been brewing in fits and starts for years.

The sequence of events began in the late 1990s, when Ecclestone was approaching his seventieth birthday. He was a decade

removed from running the Brabham team, but like all team owners, he hadn't forgotten how he spent most of his time: scrambling
around to find more money, a skill that also applied to running
the whole sport. So in order to raise a lot of cash quickly, Ecclestone had done the most fashionable thing you could do in the late
1990s that didn't involve hair dye or ill-fitting jeans. Like plenty of
other entities in the sports world, including more than two dozen
English soccer teams, he explored an IPO.

The plan was for the Salomon Brothers investment bank to
prepare an offering that valued the sport at around $4 billion for
flotation in London and New York. What that really meant was
selling all the parts of F1 that Bernie controlled under a single umbrella. There was the old FOCA, alongside a company he'd spun
off to deal in television rights called International Sportsworld
Communicators, and the arm for selling the Grands Prix themselves, known as Formula One Promotions and Administration.

The whole package was really propped up by TV, which had
ballooned through the 1990s. Ecclestone had recently sold the
British rights to ITV for nearly nine times what the BBC had been
paying. F1 claimed that each race was watched by some 330 million viewers in 130 countries, blowing away the Super Bowl every
couple of weeks.

But Salomon Brothers found in 1997 that picking the right
value for Formula 1 was trickier than planned. Bankers could sort
through tax filings and ownership documents, yet those painted
only a partial picture of everything Ecclestone controlled. The
true web of connections, rights, support structures, and ancient
deals that kept the sport turning was almost impossible to grasp.
No one could be entirely sure of what Bernie was selling. "Even
his age (somewhere above 65) is unknown," *The Economist* wrote
at the time.

That obscurity had always been to Ecclestone's benefit, until
someone started poking around with a flashlight. That someone
turned out to be the European Union, which felt that the pinnacle
of motorsport looked far too much like a monopoly to be floated
on the stock market. The arrangement between Ecclestone, the

companies he controlled, and the FIA, where Bernie was a vice president, all seemed a little too cozy.

In particular, regulators wanted to examine the agreement between the FIA and Ecclestone's Formula One Administration Ltd., which guaranteed him exclusive control of broadcasting rights for Grand Prix racing from 1997 until the end of 2010, on top of his fourteen-year lease on the sport's commercial rights. This was far too long for the EU's taste.

The dossier landed on the desk of the EU competition commissioner, who was a former leader of the Belgian Socialist Party named Karel Van Miert. When he met a reporter from the *Wall Street Journal* in 1998, Van Miert waved a folder containing details of Formula 1's structure and potential antitrust issues in the air, utterly incensed: "I've never had a case with so many infringements. And I've seen many cases.

"They thought they could come to Brussels, say they want an exemption, and then catch the plane back," he added.

Van Miert wasn't in a terribly charitable mood either. The EU was already at loggerheads with Formula 1 over its proposed ban on tobacco advertising across the continent. Ecclestone was so spooked by the prospect of losing this lifeblood of the sport that he enlisted a powerful ally to help him push back: the newly elected British prime minister, Tony Blair.

Or at least Ecclestone thought he had. Shortly before the UK election in May 1997, Bernie had made a £1 million donation to the Labour Party. The move, he said, was purely out of admiration for the bright, dynamic Labour Party leader. Ecclestone insisted that he expected nothing in return. That the EU was preparing to go to war over tobacco ads was mere coincidence.

Either way, the regulators had Bernie's attention. Tobacco was pumping roughly $250 million a year into the sport and seven of the twelve F1 teams that season had the logos of major tobacco partners slapped on their cars. These included Michael Schumacher's Ferrari, brought to you by more than $80 million from Marlboro; the Benetton and Minardi teams selling a Japanese smoke called Mild Seven; McLaren decked out in the silver and black of a German cigarette named West; and the French Ligier outfit hawking Gauloises, of course.

Ecclestone realized that any effort to boot these from the sport could cripple his entire series. Half a dozen teams would go straight into bankruptcy. His circuits would be stripped of crucial advertising dollars. And thousands of hours of television exposure would suddenly go up in smoke. So in order to preserve the relationship with the tobacco companies—and keep his sport alive—he and Mosley were prepared to take drastic action. If tobacco had to leave Europe, then Formula 1 would pack its bags too.

"If we have to move races away from Europe, then Formula 1 would continue," Ecclestone told the London *Times* in 1997. "I have been moving east for years because I think it is the biggest growth area for us and, by coincidence, it happens it is the biggest growth area for tobacco.

"It would be sad to lose the traditional circuits, places like Silverstone," he went on, claiming that 64 percent of F1's viewers in those days were based in Asia. "But the world championship is the most important thing . . . The new races would become traditional before long. You lose tradition quickly in this sport. We go forward all the time."

It was around then that Tony Blair, freshly installed at 10 Downing Street, had a change of heart that suited Ecclestone. Despite an earlier public campaign promise to support the ban on cigarette ads, his government decided that getting rid of tobacco advertising in sports would in fact be terrible news for the country. Some fifty thousand British jobs in motorsport were at stake, he said.

This reverse-ferret became particularly awkward for the government when news of Ecclestone's substantial donation hit the papers. Keen to put the scandal to rest and air out the whiff of sleaze, Blair had the Labour Party return Bernie's million. Still, his government stayed the course on its position. In Brussels, the UK continued to lobby for a Formula 1 exemption, where it eventually reached a compromise for the FIA to phase out tobacco advertising by 2006 instead of scrapping it overnight.

On the track, this deal led to a weird pantomime in places where the ban was already being enforced. Teams complied by doing the bare minimum: they removed the names of the tobacco brands from the cars, but not the fonts or the colors. For all intents and purposes, the machines still looked like cigarette packs on wheels. The McLarens of Mika Häkkinen and David Coulthard remained silver and black, except the lettering on the side of the car read "Mika" and "David" instead of "West." And at one Grand Prix, Ferrari simply replaced the word "Marlboro" on its rear wing with the Formula 1 logo. The Marlboro cowboy might as well have been winking straight into the camera.

With such brazen behavior on F1's part, the case that landed on Van Miert's desk in Brussels in 1998 was never likely to break in the sport's favor. In short, when the European Union looked at Formula 1 racing, it saw an anticompetitive, pro-smoking, highly secretive operation run by some blazers in Paris who were far too close to Ecclestone, a man who paid himself $80 million in 1996 and, in the EU's view, treated prime ministers like personal lobbyists.

Bernie thought the EU couldn't be more wrong.

"I feel sorry for the commissioners," he said at the time. "Half the people in Formula 1 don't understand what goes on in our business."

Nor did Ecclestone really want them to find out. What was clear, however, was that his plan to raise cash by taking Formula 1 public was dead in the water. If he was going to properly brace the sport against the loss of tobacco money down the road, it was time to consider other options. The next best thing he could come up with was to get into the debt business. His plan came to be known as the Bernie Bond.

Ecclestone initially engaged Morgan Stanley in 1998 to handle the bond issue and liked what he was hearing from the bank when it promised a valuation of $2 billion. The problem was that no one would want to go near it. Formula One Administration, the principal company in Ecclestone's empire, had shown operating profits of £76.5 million in 1997, which was a long way off what was required to build confidence in a $2 billion Bernie Bond.

As it turned out, only one person was willing to say so. She was an American named Robin Saunders, working for a sleepy regional bank in Germany called West LB, which had been brought in by Morgan Stanley. Until meeting Bernie Ecclestone in person, Saunders barely had any idea who he was. But one thing she knew for certain was that his numbers didn't add up.

Sitting in a too-crowded, too-hot little conference room that staffers were sure was bugged at F1's London headquarters in Princes Gate, Saunders told Ecclestone point-blank. The valuation would have to come down to £1.4 billion and he would need to guarantee some degree of transparency. Otherwise, this could never work.

"Everybody was scared of Ecclestone," she says. "No one knew if he was a genius or a crook."

The due diligence process didn't shed much light on that question. Ecclestone was so secretive that he wouldn't let a single document leave Princes Gate and photocopies were banned. Whatever snooping around the lawyers and accountants wanted to do had to be done in the tiny conference rooms with the heat cranked up.

Because so much of Formula 1's institutional knowledge rested with Ecclestone—and no one else—any deal also needed to include provisions for what businesses call a "key man." In other words, Bernie was essential to the health of the enterprise, so investors needed to be sure that this soon-to-be seventy-year-old wasn't likely to (a) retire or (b) drop dead anytime soon. Morgan Stanley assured Saunders that all of those health checks had been made. And Ecclestone, despite his impatience over the course of six long months, was surprisingly receptive to Saunders's most compelling argument throughout the process.

"If you want to become a billionaire for the first time in your life," she told him, "this is what you've got to do."

The deal finally came together in early 1999, and on one spring Friday, the money was deposited into Bernie Ecclestone's empire, though not directly into Bernie's account. Technically, it went to a company called SLEC Holdings that didn't belong to Ecclestone at all, but rather to the six-foot-one Croatian former model whom he'd married in 1985. At the stroke of a pen, Slavica Ecclestone became Britain's second-wealthiest woman, behind only the Queen.

Saunders had pulled it off. The British papers would later call her the "Fairy Godmother of Formula 1." But first, there was one more freakout.

On Monday morning, just days after the bond issue, Ecclestone rang Saunders with a confession, which struck her as strange, since he wasn't in the habit of being especially open with anyone. In six months of working on the project, the pair had never even had a social cup of coffee.

"Robin," Ecclestone said, "I've been really naughty."

Saunders started to panic. *Oh my God, this is what everybody was always worried about. He really is a crook,* she told herself, *and now I'm about to get the big reveal.*

"I actually," he said, "have to have a triple bypass."

Saunders was floored and apoplectic all at once. The whole bond issue hinged on a promise that Bernie would be around for a while. She was certain that Morgan Stanley had told her that there had been health checks. Ecclestone had assured her, too, that there was nothing to worry about. Now Saunders knew that her whole career was on the line because this old man had lied to her by omission. There was no Plan B if something went wrong.

"Okay," she said, concealing how livid she was. "When?"

"Two o'clock," Bernie told her.

"If you make it out alive," Saunders replied, "I'm going to kill you myself."

ECCLESTONE SURVIVED THE SURGERY JUST FINE—"I DISAP-pointed so many people," he says. And instead of murdering him, Saunders accepted a spot on the company's board.

That gave her a front-row seat to a frenzy of activity that would see the business of Formula 1 carved up, sold far and wide, and ultimately reconstituted, all in the space of five years. There were private equity firms, investment banks, and even a German media conglomerate named EM.TV, which had recently bought the rights to *The Muppet Show*. Some lost their shirts investing in F1. And some, like the US-based firm Hellman & Friedman, made around half a billion dollars on the sport despite owning a stake for barely six months.

But when the dust settled in the early 2000s, 75 percent of the business had been reconsolidated in the hands of one man, an EM.TV backer named Leo Kirch. Just as Michael Schumacher was finding out on the track, it was a good time to be a German in Formula 1.

Kirch, who was conscripted into the Luftwaffe as a teenager and then deserted just before the end of the Second World War, laid the first bricks of his empire by acquiring foreign film rights to distribute in Germany. He soon built himself into a figure not unlike Rupert Murdoch and Silvio Berlusconi in the country's media landscape. In fact, he followed the Murdoch and Berlusconi models so closely by betting on pay TV and soccer broadcasts that he eventually became Murdoch's business partner in the venture that is now Sky Deutschland. The difference there was that Germany didn't care much for pay TV.

In 2002, KirchMedia found itself 6.5 billion euros in the hole. The largest shareholder in Formula 1 was quickly going bust. And when it slumped inevitably into bankruptcy, the company's creditors were staring at a pile of rubble that just happened to include the world's most popular racing series. After passing through the hands of Bernie Ecclestone, Slavica Ecclestone, and a collection of global investors, most of F1 now belonged to a group of banks that had no idea what to do with it.

None of them had any meaningful control over the sport, but the idea of reporting to Bayerische Landesbank, JP Morgan, and Lehman Brothers irked Bernie no end. Every board meeting seemed to turn into an expletive-ridden shouting match.

"I had a lot of bosses," Ecclestone says.

That's when he turned to CVC. The banks were desperate to recoup their money and Bernie was desperate for some different bosses. CVC was certainly intrigued, even if Ecclestone didn't paint the rosiest picture. He told them he was at war with the banks, that the business was going down the tubes, and that the teams were so pissed off that they were threatening to leave Formula 1 and start their own racing series. "The company was a zombie company from 2001 to 2005," says CVC's Nick Clarry, who worked on F1 for twelve years. "They just wanted to sell the shares to anyone who walks in the door to get their money back and repay the loans because they were lenders."

The issue for CVC, which also had to sell off Moto GP to comply with the European Commission's competition rules, was securing the funding. F1 was so toxic that no one wanted to touch it. Only the Royal Bank of Scotland came through, in part because it was already a sponsor of the Williams team. CVC finally structured a deal to acquire the F1 Group in a leveraged buyout for $2 billion in 2006. They just had one more query for Ecclestone before signing the paperwork. Donald Mackenzie asked him one last time whether there was anything he should know about the company that he wasn't aware of already. He didn't want another surprise bypass announcement.

"I'll tell you what I'll do," Ecclestone said to him. "I'll give you a £100 million guarantee that there are no things that are gonna disturb the company."

That seemed good enough to CVC. Besides, Bernie told Mackenzie he was ready to go back to work: "Can we get on with it?"

FORMULA 1'S NEW OWNERS SOON FOUND THAT FORMULA 1'S old owner was worth keeping around. He didn't like being told what to do, and he was capable of generating the odd catastrophic headline, but his business style, his relationships, and the address book in his flip phone meant Bernie Ecclestone could still be very useful—especially when it came to the F1 calendar.

Races fell into three categories. There were the traditional hosts, like Monaco and Monza, getting by on history and prestige. There was a middle tier of long-standing but easily rotated circuits paying north of $20 million for the privilege—think Australia or Canada. And then came the new Grands Prix in the markets F1 coveted most: China, the Emirates, anything east of Budapest. "Then you paid top dollar," says Michael Payne, a veteran of Olympic sponsorship and broadcasting who came in to assist Ecclestone.

Those top-dollar tracks were Bernie's bread and butter. CVC had said from the outset that it wanted to keep growing Formula 1 globally and take it into new markets. The challenge, as ever, was keeping the number of races reasonable. During CVC's first season, there were eighteen Grands Prix on the calendar. Any more than that and the teams would have been out for blood. Why burn out their staff and run up their expenses if they still weren't getting even half the cake?

But on closer inspection, CVC and Ecclestone found there were plenty of ways to rebalance the mix of circuits without necessarily adding dates. They just needed to trim some fat. In 2006, the schedule featured two races in Italy, two in Germany, one in France, one in Spain, plus Great Britain and Monaco. "It was a very Eurocentric business," Clarry says.

Ecclestone had already managed to add China for 2004 and Turkey for 2005. The year CVC took over, he held his first meeting with the organizers of Abu Dhabi and had designs on a race in India. But there were still bigger prizes to come. The first was Singapore, where Ecclestone had dreamt of holding a race for at least a decade. He got his wish in 2008 with a race held at night to suit the European audience, under the lights, around the streets of Marina Bay. The cost to the promoters, paid in part by the government of Singapore, was over $100 million.

"The only way you're going to create a successful global franchise is through a little bit of benign dictatorship," says Payne.

There was still one venue that Bernie wanted even more. He'd been fantasizing about holding a Grand Prix in Russia for so long, in fact, that when he started trying to make it happen, the country was still called the Soviet Union. The proposal never got off the ground and Ecclestone achieved his goal of taking Formula 1 behind the Iron Curtain in Budapest instead.

But by the late 2000s, Russia had reentered the global sports frame. The country was preparing to host the 2014 Winter Olympics in Sochi at a cost of more than $50 billion. Then four years later, Russia had the rights to host the soccer World Cup. With all of those rubles flowing into global sports, F1 could smell its chance.

As it happened, Payne had plenty of connections in the Russian Olympic world from his years at the International Olympic Committee. When he approached them, he asked what their plans were for the city and the region after the Games: "How are you going to keep your tourist region front and center? How are you going to build the brand with Sochi?" Formula 1, he explained, was an opportunity to put Sochi, a resort town on the Black Sea, in a club of luxurious, globally recognizable locations alongside Monaco and Singapore. The so-called Caucasian Riviera would have a legitimate connection to the actual Riviera.

After months of work, Payne had the brass in Sochi convinced that they needed a Grand Prix. That turned out to be the easy part. Next, he had to convince Bernie that his dreams of a Russian race could finally land, but only on the shores of the Black Sea.

"I don't want to go to Sochi—it's a dump," Ecclestone told him. "Why would I want to go there? I want to go to Moscow or St. Petersburg."

"Forgive me," Payne replied. "You've been trying to go to Moscow and St. Petersburg for twenty or thirty years. There's a very good reason you haven't gone: because they won't pay you. If you want to get paid, we're going to Sochi. And I would argue that $50 billion later, it might not be such a dump."

Ecclestone wasn't so sure. It didn't help his impression much when a delegation of officials from Sochi failed to turn up after he'd invited them to be his guests at the Monaco Grand Prix. Ecclestone had grown so cool on the idea that Payne all but gave up. Only when he walked into Princes Gate one day in 2010 did Bernie inform him that he'd been invited to go to Sochi the next morning to meet personally with President Vladimir Putin for a discussion about Formula 1. But, Bernie added, he didn't feel like going.

"Why not?" Payne asked him.

"Do I look stupid?" Ecclestone answered. "If they think I'm going to go and negotiate this contract with Putin, no way. If they want me to go, then they can send me back the contract—signed."

Payne pointed out that the contract was two hundred pages long and worth several hundred million dollars. Perhaps a little discussion might be required. But Ecclestone didn't want to hear it. Putin or no Putin, he wasn't getting on a plane without a guarantee that Sochi would pay for a race.

Two hours later, the contract came back, Payne says, "with not a comma changed."

Within four years, Formula 1 cars were tearing around the Sochi Olympic Park and Lewis Hamilton was on the top step of the podium, spraying champagne at Putin.

BERNIE'S ROLODEX WAS THE UPSIDE. BUT IN 2006, CVC'S NEW-est employee also created his fair share of headaches.

Though the various businesses under the F1 umbrella were pulling in around $1 billion a year, only $250 million of that was finding its way to the teams. All of them blamed Ecclestone and were beginning to whisper that they might break away, possibly backed by cash from Rupert Murdoch. The way F1 looked on-screen badly needed updating, too, since most races were still broadcast by the local partners who focused on hometown drivers and simply beamed out those pictures to the rest of the world. Then there was the small matter of figuring out what to do about developing F1's presence in media more modern than cathode-ray television. Ecclestone freely admitted that he didn't really understand the internet—and didn't have any interest in learning about it either.

Even so, CVC found him reasonable to deal with in the early days. If nothing else, he kept things entertaining. When Bernie felt a race weekend was too quiet, he and Mosley liked to summon a journalist or two to feed them a spicy line to earn Formula 1 a bit more real estate in the next day's papers—a test balloon for a rule change, a rumor about cutting a Grand Prix, anything to stir the pot. (And that was before Ecclestone's 2009 comments praising Adolf Hitler as a "man who got things done.")

"Did we love all of the quotes and interviews Bernie gave the newspapers? No, not really, because, you know, he likes to court controversy," Clarry says. "But he was an entrepreneurial genius. And we put an infrastructure around him to institutionalize the company."

That included a board of directors over him for the first time, which he had no choice but to accept. On some things, though— well, on a lot of things—Ecclestone could be intransigent. The issue of succession, for instance, was a topic he refused to touch. CVC wasn't accustomed to betting billions of dollars on the health and mental capacity of a septuagenarian. Yet it knew that it could never replace Bernie with anyone else, because Bernie was really half a dozen jobs packaged under a mop of silver hair. Anytime CVC tried to hire talent to put alongside him, Ecclestone steadfastly blocked it.

"Succession, yes, yes, yes," he would tell his new bosses. "I totally understand how important it is. And twenty to thirty years from now, I'll be very happy to sit down and talk about succession."

CVC didn't plan to hang around that long. Their model was to be in a business for five or ten years, doll up the operations, and turn a tidy profit on the sale. Signing a new Concorde Agreement to keep the peace between the teams, the commercial rights holder, and the FIA for 2007 seemed to help. But the closer CVC looked at the inscrutable business of Formula 1, the more complex it became.

The tobacco question still hadn't been solved and the sponsors were being squeezed out of the game. Team spending, meanwhile, was spiraling like the brakes had malfunctioned. And years of

Ferrari dominance had dented the ratings. Yet CVC felt that all of those issues were fixable with smart management and business practices that were a little more sophisticated than selling used cars in the 1960s.

What CVC wasn't prepared for was that this cutthroat business didn't just play out on conference calls, in fancy motor homes, and in the cramped meeting rooms of Princes Gate. This was also a vicious, high-stakes competition between millionaire sharks who would stop at nothing to find an edge.

The sport's new owners learned that the hard way. Less than two years after buying Formula 1, the whole enterprise was soon on the verge of unraveling due to a shitstorm they never saw coming: one of the most flagrant, most expensive cheating scandals in sports history, immediately followed by an even more flagrant and expensive cheating scandal.

And it all started in a village copy shop.

9

Spygate

ONE MORNING IN EARLY JUNE 2007, Trudy Coughlan walked into an unremarkable Surrey storefront with an unwieldy stack of 780 pages that she wanted transferred onto CD-ROMs. Gary Monteith, who ran the copy shop, didn't usually pay much attention to what was on his customers' pages—it was none of his business. But as he ran his thumb over the ream of paper, something printed on the sheets jumped out at him.

Everywhere he looked, he saw the black Prancing Horse of Ferrari. Monteith had supported the Scuderia as an F1 fan for years, and if all that time had taught him anything, it was that Walton-on-Thames, Surrey, sat an awfully long way from Maranello. Sensing an opportunity for some Formula 1 chitchat, Monteith asked Coughlan, "Do you work for Ferrari?"

"Err . . . my husband does," she stammered. "Well, he used to."

The part she left out was the identity of her husband's current employer: Mike Coughlan was actually the chief designer at McLaren. The closest he'd come to wearing the red uniform of Ferrari was when he worked as an assistant under John Barnard, the designer who'd refused to move to Italy. But ever since 2002, he'd been based in Woking as one of Ron Dennis's most trusted lieutenants.

This is what Monteith discovered for himself through some light investigation the moment Trudy Coughlan had left his store. He'd typed her name into a search engine and a few things suddenly fell into place. It made perfect sense that the man she was married to worked for McLaren. The team's HQ was only ten miles down the road.

What didn't add up was why this McLaren guy and his wife were in possession of detailed specifications for that year's Ferrari. Something wasn't right. People didn't just waltz into his copy shop with 780 pages of highly confidential materials—they brought utility bills and birth certificates. So Monteith decided that he needed to tell someone. But who? As a *tifoso*, he knew the name of Ferrari's sporting director, Stefano Domenicali, from the TV every other Sunday. And it didn't take long for Monteith to work out an email address for him.

"Dear Mr. Domenicali," he began. "My name is Gary Monteith. You don't know me, but I'm a big Ferrari fan and there's something I thought you'd want to know . . ."

Monteith's suspicion was that the 780 pages now in his possession were in fact private property of the Scuderia that had somehow fallen into enemy hands. If that were true, he could scarcely imagine the consequences. Hundreds of millions of dollars had been spent to develop the contents of this document. Every page was potentially covered in trade secrets. As far as he could guess, this might even be a criminal matter.

Hoping to grab Domenicali's attention, Monteith laid out his fears in the starkest possible terms, right there in capital letters in the subject line. Before hitting send, he typed out the two words that would change the course of F1 history.

"INDUSTRIAL ESPIONAGE."

THE MOST UNLIKELY PART OF THE WHOLE INSANE EPISODE soon known as Spygate was that Stefano Domenicali not only received Monteith's email but actually opened it. Domenicali, who might be the least impulsive Italian ever to roll through the doors of the factory, was the consummate company man. Anything that could be seen as a threat to Ferrari felt like a personal threat to Stefano.

Born in Imola, less than sixty miles from Maranello, Domenicali grew up steeped in the legend of the Scuderia. He studied at the University of Bologna, which might as well have been the Ferrari Institute of Technology, and then joined the company as soon as he could. He scaled the ranks through the sports department, with a brief detour as logistics manager. And now, nearly two decades later, he understood Ferrari as well as anyone who wasn't related to Enzo. So when he read Monteith's email, Domenicali knew instinctively what the correct procedure was. He referred the email to Ferrari security, looped in the legal team, and forwarded a copy directly to Ferrari chairman Luca di Montezemolo. Someone needed to check this British guy out.

"I was furious," Montezemolo says. "Thank God, we have Ferrari fans all over the world—even in the UK."

Beyond its fans in the UK, Ferrari now needed legal help there too. The man tasked with working out whether or not Monteith was a crank was a City of London lawyer named Duncan Aldred, who specialized in corporate fraud. The question before him was whether Monteith had uncovered an act of fraud or was merely in the process of committing one. After all, this was the era when British tabloid reporters liberally went undercover dressed as an Arab sheikh or a Buckingham Palace footman to trap people into major exposés.

Aldred invited Monteith to his office, and quickly surmised that this wasn't a man capable of orchestrating an elaborate plan to scam Ferrari. From the moment he walked into the law firm sporting a backpack, Monteith struck him as a fish out of water among the pinstripe suits and briefcases. Aldred got straight to the point. When he posed specific questions about which details of the documents Monteith could remember, the man from the copy shop simply picked up his backpack and poured out the contents on the conference table. Out spilled 780 pages of highly confidential Ferrari materials. Monteith had made a copy for himself.

The gravity of the situation became immediately clear to Aldred. This wasn't just the F1 equivalent of an NFL team obtaining a rival's playbook. This was like an NFL team getting its hands on the playbook, the payroll, the draft board, and the medical records of every player on the roster. All of the design specifications of that year's Ferrari F1 car were laid out in this bible of engineering and someone at McLaren had been flipping through it like it was that morning's newspaper. Aldred sensed immediately that the documents were genuine, and years of experience told him that this represented a matter of the highest urgency.

He told Ferrari that the company had a decision to make. Turning this into a criminal case risked dragging things out pending a full police investigation while McLaren continued about its business. But the British legal system offered another recourse, Aldred explained. They could obtain a civil court order that would allow Ferrari lawyers to search Mike Coughlan's home as soon as possible and recover any proprietary materials they found there.

Ferrari chose the latter. And at 7:30 a.m. on the morning of July 3, Aldred and a team of attorneys descended on Coughlan's house near Woking. The surprise visit caught the forty-eight-year-old McLaren engineer just as he was folding his burly frame into a team-issue gray Smart Car.

Confronted with the facts, Coughlan didn't put up much of a fight and realized that his best course of action was to start talking—it's not as if he would have gotten very far in his tiny electric vehicle anyway.

Yes, Coughlan told them, he had been in possession of the 780 pages. And yes, he had also sent Trudy to have them digitized. But he was at pains to explain that he hadn't obtained them illegally or through deceit. Besides, all that existed were the CDs. The original hard copy had been shredded and burned. He'd been given them directly by one of Ferrari's highest-ranking mechanics, a fellow Englishman and former colleague at Lotus and Benetton named Nigel Stepney. The way Coughlan told it, Stepney had harassed him for months, desperately trying to leak the Ferrari secrets to him. Precisely what motivated Stepney to do this was never quite clear, but Coughlan did know that Stepney wasn't the most popular employee of the Scuderia.

A longtime deputy of Ross Brawn's, Stepney had joined Ferrari in the early 1990s and risen to the rank of chief mechanic. Tall, with neatly cropped hair and a goatee, he had learned to speak Italian, moved to Maranello, and was directly credited by Brawn with Ferrari's outstanding reliability throughout the Michael Schumacher dynasty. But Stepney could also be crude and a tyrant to his underlings. The mechanics inside the Scuderia—the foot soldiers who form a powerful constituency within the team and view themselves as the true guardians of the Ferrari soul—never quite took to him. That turned out to be a fatal blow to Stepney's ambition to become the team's technical director after Brawn left at the end of 2006. Team principal Jean Todt sensed it too.

"If you're a No. 2," he told Ferrari insiders, "you'll always be a No. 2."

Stepney, who had devoted fifteen of the best years of his career to the team, felt betrayed. And now it seemed, in true Italian operatic style, he had turned that betrayal into a vendetta. At least that's how Coughlan portrayed it. The months of phone calls, he said, had become a nuisance. Stepney invited him to meet during preseason testing in Barcelona in early 2007 and Coughlan went along—but claimed that he only accepted the meeting to tell Stepney to stop contacting him. As it turned out, this was believed to be the rendezvous where Stepney turned over the Ferrari documents.

Anyone who heard Coughlan's version agreed that it was a ludicrous argument. When he heard about the raid, the documents, and the excuses, McLaren team boss Ron Dennis was rattled. He discreetly suspended Coughlan just days before the British Grand Prix.

The matter wouldn't stay quiet for long. That same week, the FIA announced that it was opening an investigation, just as Dennis was preparing to unveil his latest pride and joy at Silverstone. It wasn't a piece of state-of-the-art engineering, but rather his latest innovation in how to make money: the McLaren "Brand Centre," a space-age-gray headquarters-on-wheels where he could make sponsors feel that they were part of the slickest, most advanced, most meticulous team on the paddock. Only now, as Dennis scrambled to do damage control with guests and journalists about to arrive, he was freaking out over one detail of the Brand Centre that he'd overlooked. In the midst of a burgeoning espionage scandal, Dennis ordered McLaren staff to remove the bottles of white wine that they were getting ready to serve. The sauvignon blanc was from a place in New Zealand called Spy Valley.

Dennis spent the weekend brushing off the questions as best he could. McLaren's position was that any wrongdoing was the work of a single rogue employee. In no way did Coughlan have any backing from the team in his dealings with Stepney—whatever they were—nor did McLaren benefit from his actions. Dennis explained that he and his team were "not perpetrators, but victims, in the same way as Ferrari is."

Over at the FIA, Max Mosley's trained legal instincts felt that this might require some closer examination. Years earlier, Mosley had referred to Dennis as "not perhaps the sharpest knife in the box"—and he certainly wasn't a lawyer. So in matters of high importance with enormous repercussions for the image and integrity of the sport, Mosley wasn't exactly inclined to trust his homework. On July 12, Mosley summoned McLaren to answer a charge of unauthorized possession of documents at an "extraordinary meeting" of the World Motor Sport Council at the Place de la Concorde in two weeks.

That was scarcely enough time for Dennis to assemble a team of lawyers, but he knew how high the stakes were. For decades, McLaren had prided itself as the team that showed the rest of the paddock how things should be done, forever holding itself to the highest standards Dennis's obsessive mind could imagine. It was the first to paint the garage floor, the first to treat every area where mechanics worked on the car with the seriousness and cleanliness of an operating theater. The team was designed not only for speed, but for spotless presentation—this had been Ron Dennis's life's work. Now he was staring at the one stain he couldn't scrub out. The cheating allegation against what he believed was a single employee acting alone had mushroomed into an attack on his entire personal code.

On July 26, Dennis arrived in Paris wearing a dark suit with a gray tie feeling like he'd experienced twenty-four days of hell. He was determined to clear the McLaren name and, more important, the Ron Dennis name. "I am an honorable person," he said at the time. "My company has been dragged through filth, through spin and innuendo."

More than two dozen officials from the FIA, McLaren, and Ferrari crammed themselves around the horseshoe table at the world governing body headquarters to weigh that claim. This being Paris, the air-conditioning was functioning at less than optimal capacity. But presiding over this crowded, sticky de facto court-room, Mosley was in his element. The conversation quickly devolved into a one-on-one interrogation with Dennis on the stand and Mosley operating as prosecutor, judge, and several members of the jury.

Under questioning, Dennis did his best to stick to McLaren's version of events: Coughlan was a rogue employee, who acted alone, and never fed any of the Ferrari information he obtained from Stepney into the team. The whole thing, he argued, was a plan cooked up by Stepney to get them both a big money offer to join Honda.

"We have 136 engineers and designers, totally focused on chassis development—one bad apple is immediately apparent," Dennis said. "We have completely verified, for our own standards and our own satisfaction, that there were no Ferrari drawings or ideas on our premises. Coughlan did not communicate into the system anything with origins in Ferrari intellectual property."

"Did he contribute anything?" Mosley inquired.

"His contribution . . . his contribution is the sign-off," Dennis sputtered. "His peak workload comes in November–December–January, when ten to fifteen thousand drawings come in. Someone must deem them 'fit for production,' or 'under consideration for drawing,' 'under consideration for structural properties,' etc. That is not the design process."

Mosley let the answer hang in the air. Dennis appeared to be suggesting that McLaren's chief designer was not actually involved in the design process.

Then Mosley pounced. "You are thus saying that he made no contribution?"

Dennis hesitated. "There is not a single employee at McLaren who does not contribute to our winning races," he shot back.

Mosley was clearly toying with him. He pointedly addressed Ferrari's Jean Todt as "Mr. Todt," but called Dennis "Ron." And Dennis wasn't the only one squirming. Even other people in the room were finding the whole thing a little bit uncomfortable. For all of Dennis's masterminding of a successful Formula 1 team, he wasn't equipped with "a massive vocabulary," as one person in the room remembers. "Not really a thinker in the same way as Max Mosley [who was] enjoying making a bit of a fool out of him."

As the hearing went on, Dennis grew more and more frustrated, irked by the way Mosley seemed to be talking down to McLaren employees.

"I watch him being intimidated and do not like it," Dennis piped up after one particularly tense exchange.

"Ron—" Mosley began.

"I am just being defensive," Dennis interrupted, not quite using the word properly.

"Please do not think that we are intimidating anybody."

"I feel intimidated!"

"He is not intimidated," Mosley informed Dennis. "You may be, but he is not."

The proceedings went on this way for hours. But when they ended, McLaren got away with a slap on the wrist. The FIA found that the team was, in fact, in possession of Ferrari's intellectual property, but no one had established beyond a reasonable doubt that McLaren's engineers had used it for anything.

"Although no one on the council believed them, we had to acquit," Mosley wrote later. "Concrete evidence of use by McLaren of the Ferrari information was simply not there. Without it, McLaren would have won an appeal to the FIA International Court of Appeal."

Ferrari was incensed by the verdict, and Fiat CEO Sergio Marchionne rang Mosley in a rage to let him know. "Without evidence, we could not convict," Mosley repeated, "however much we might believe them to be guilty."

Dennis, for his part, had every reason to be elated. Facing a possible ban from Formula 1 with all of the ripple effects it would have on his McLaren business empire, he'd escaped any serious punishment at all. There was just one thing sticking in his craw. Would the FIA mind removing the word "guilty" from its press release and any further communiqués that described the team's illegal holding of Ferrari materials? He worried that the G-word would reflect negatively on the McLaren brand.

Mosley didn't appreciate Dennis's temerity and proceeded to ignore his request. In closing the hearing, he issued a warning: should further evidence of McLaren wrongdoing later emerge, Mosley reserved the right to reconvene the World Motor Sport Council and go through the whole song and dance again. Dennis would have to sit in the hot seat for another grilling.

"It is thus very important that, if there is anything else we ought to know, we be told it [now]," Mosley cautioned Dennis and the McLaren attorneys. "Clearly, it is much worse if such matters are unturned later."

Those matters could be anything, Mosley added. A detail like the weight distribution of the Ferrari car, for instance, definitely seemed like something that would merit internal discussion at McLaren if that information happened to be in the documents. Just as an example.

McLaren kept quiet. But Mosley wasn't the kind of man to say things idly, especially not about obscure-sounding technical specs like weight distribution. With informers throughout the world of F1, he seemed to know something that even Dennis might not. The hearing was over, yet Mosley seemed to be suggesting that Spygate wasn't.

"I still think there was a bit of the jigsaw puzzle that only Max Mosley knew," one witness says.

FERRARI VS. MCLAREN WASN'T JUST A BATTLE INSIDE FIA headquarters. It was also a knife fight on the track.

With or without supplementary reading material, McLaren had clearly produced the quickest car of the 2007 season. Plus, it had the undisputed quickest driver in the sport: Fernando Alonso, who was coming off back-to-back world championships with Benetton, seemed set to replace F1's years of German dominance with a Spanish empire. He was going up against Ferrari's Kimi Räikkönen, a somber Finn who'd nearly been fired during his four years as a McLaren driver for failing to do things the Ron Dennis way.

But over the early part of the season, it became clear that Alonso was facing an even greater threat than Räikkönen—and it was coming from inside his own team. A young British rookie named Lewis Hamilton had finished on the podium in each of his first nine F1 races. Hamilton didn't view himself as McLaren's No. 2 driver. And he underlined it at the opening corner of the opening Grand Prix of the year in Australia when he shot past his teammate with a daring move around the outside.

Dennis could tell then and there that very soon, this was going to become his problem. The man who'd been in charge of refereeing the blood feud between Senna and Prost had inadvertently created another internal power struggle. Hamilton was too young and too cocky to know any better. But Alonso still took it upon himself to teach his junior teammate a series of lessons.

The ugliest incident came in Hungary, where Alonso deliberately impeded Hamilton by parking in the pit lane during qualifying and preventing his McLaren teammate from completing his final lap as payback for what Alonso took as a lèse-majesté earlier in the day. Hamilton had shown the impudence to disobey team orders and drive his car out of the garage and onto the track first. No one who understood the delicate internal hierarchy of an F1 team would ever do such a thing. Hamilton didn't care.

The situation had already been festering for months. Now it was going bad faster than a tuna sandwich in the sun. On the Saturday afternoon after qualifying, where Alonso had secured pole position, Dennis tried to smooth things over. But the Spaniard had no interest in playing nice anymore. As if to antagonize his obsessive-compulsive boss on purpose, he sat in a media briefing next to Dennis while breaking one of Ron's cardinal rules: he was ostentatiously eating an enormous, succulent peach with his bare hands. For Dennis, the sight of someone consuming a piece of dribbly fruit without using a knife and fork to cut it into tidy pieces was almost too much to bear.

By the next morning, Alonso was even more furious, because the stewards had handed him a five-place grid penalty for the race following his shenanigans in the pit lane, dropping him from pole position to sixth. Feeling like his team was turning against him, he demanded that McLaren do something to remind Hamilton of the pecking order.

The idea Alonso had in mind was not giving him enough fuel to finish the race. Dennis dismissed that absurd suggestion out of hand. But Alonso wasn't kidding. He showed Dennis just how deadly serious he was by hitting him with the most vicious threat he could think of: Alonso brought up Spygate. Days after Dennis had escaped the first hearing largely unscathed, Alonso said he had evidence that the contents of the Ferrari dossier had actually circulated far and wide within the McLaren team. The drivers were implicated too—and he had the emails to prove it. Unless Dennis put Hamilton in his place, Alonso would break the whole thing open.

Even Senna had never gone this far.

Before the race, which would see Hamilton storm to victory and open up a 7-point lead in the championship standings, Dennis was in a panic. If what Alonso said was true, McLaren faced the prospect of being banned from Formula 1, perhaps permanently. His entire professional legacy was on the line.

Dennis felt he had no choice but to pick up the phone and call the last person in the world he ever wanted to speak to about this: Max Mosley. In all likelihood, Mosley was already aware of the emails. But less than two weeks removed from the Paris hearing, he knew that they would have to get back into FIA headquarters and do it all again. Mosley convened a second extraordinary meeting of the World Motor Sport Council for September 13. And just to make sure that everyone heard what Alonso had to say, he granted amnesty to any driver who came forward with testimony. Dennis suddenly had a lot more to worry about than the phrasing of a press release.

FOR THE SECOND TIME IN TWO MONTHS, RON DENNIS FOUND himself pushing through a crowd of photographers toward the neoclassical facade of the Hôtel de Crillon on the Place de la Concorde. Back during the French Revolution, the vast square had been the site where Louis XVI and Marie Antoinette took their final steps toward the guillotine. To Ron Dennis, walking into another session before Max Mosley and the World Motor Sport Council felt even worse.

McLaren's guilt was no longer much of a debate. Dennis was there to save his job and make sure his team wasn't decapitated.

In between the two hearings, the true extent of Coughlan's information sharing had become overwhelmingly clear and mushroomed into a public scandal that was overshadowing the season. Italian police, dragged into the affair by Ferrari, had turned up 323 text messages between Coughlan and Stepney. Over a terabyte of data had also yielded a ream of damning emails that connected the stolen dossier to Fernando Alonso and to McLaren's test driver Pedro de la Rosa.

In one short note, dated March 21, De la Rosa had written to Coughlan directly: "Hi Mike, do you know the red car's weight distribution? It would be important for us to know so that we could try it in the simulator." (There it was again—the weight distribution question that Mosley had perhaps not so accidentally dangled in front of Dennis in July.) No one needed to clarify which "red car" it was referring to. Coughlan replied by text message with the precise details of the Ferrari setup.

Another cache of emails turned up an exchange between Stepney and Coughlan under the subject line "Drag," about the aerodynamic specifications of Ferrari's front floor. Nothing was too detailed to ask about. When De la Rosa pressed Coughlan for more on the Ferrari braking system, Coughlan believed he had so much technical information available that it might actually be beyond De la Rosa's comprehension. "It may be difficult for you to understand," he wrote.

But De la Rosa wasn't asking for himself. "Fernando wants to know," he wrote back.

Once he passed it on, De la Rosa could assure Alonso that the data was solid gold. "All of the information from Ferrari is very reliable," he wrote. "It comes from Nigel Stepney, their former chief mechanic. I don't know what post he holds now. He is the same person who told us in Australia that Kimi [Räikkönen] was stopping in Lap 18. He is very friendly with Mike Coughlan, our chief designer, and he told him that."

So much for two rogue employees. That Räikkönen actually went into the pits on Lap 19 wasn't much of a defense for McLaren, which also tried to argue that because Hamilton received none of these technical specifications, only a handful of people inside the team had actually been made aware of them.

Mosley was less combative the second time around. He knew that he had McLaren dead to rights. The only question was what to do about it.

"They were clearly guilty," he wrote. "Armed secretly with the entire intellectual property of their main rivals, not to mention a flow of additional information from Nigel Stepney, their 'mole' inside Ferrari, McLaren had clearly enjoyed a massive but wholly illegitimate advantage. This called for a sporting penalty."

Suspending McLaren from the 2007 championship made the most sense. Ron Dennis and his team had the constructors' title sewn up, but there was no way they could be allowed to claim it.

The other problem was that it was already September—McLaren would have been deep into development of the 2008 car. So on top of 2007, that was another potentially illegitimate season, even if McLaren argued that no one could point to a single part it had copied from Ferrari. That defense didn't wash with the Scuderia.

"In athletics, if a runner takes a banned substance, that runner is disqualified," Ferrari's lawyer told the council. "It is not necessary to show that the runner has gained an advantage. It is enough that he has taken a banned substance."

Mosley agreed. F1 had always tolerated some degree of light plagiarism. Teams studied each other, borrowed ideas, and outright lifted any design loophole that conferred a potential benefit. This, however, had gone too far. The sheer brazenness of Stepney and Coughlan's actions had crossed a line that no one in F1 could quite define, yet everyone sensed instinctively. It was one thing to bend the rules. But in a sport literally named after the rulebook, you couldn't pretend that the rules didn't matter.

Even so, Mosley continued to fret over the impact that wiping out two seasons of income might have not on Ron Dennis—punishing Dennis was the point—but on the thousand people he employed. He suddenly pictured mass layoffs in Woking. "Their situation would have been dire," he wrote. Still, when the council deliberated over a sentence, Mosley pressed for a ban, only to be outvoted.

The solution they came up with instead was the largest fine in the history of professional sports: $100 million.

By comparison, when the NFL found the New England Patriots guilty of their own Spygate scandal *on the very same day,* the league's massive, unprecedented punishment was a $250,000 fine for the team, $500,000 for coach Bill Belichick, and the loss of a first-round draft pick. In Formula 1, those fines are about equivalent to the cost of a few steering wheels and a rear wing. Even by the absurd standards of F1 spending, the magnitude of the McLaren fine was truly staggering. It would have crippled almost any other team. Though much of the penalty took the form of withheld prize money, Dennis still had to cut a check for more than $40 million (which the team unsuccessfully tried to write off as a tax deduction).

Mosley still thought McLaren had gotten off lightly, mostly because of Dennis's histrionics and obfuscation over the course of the two hearings. As Bernie Ecclestone joked to Max in private, the breakdown of the enormous fine was actually $5 million for the infraction, and $95 million for Ron "being a c—."

ADDING INSULT TO INJURY TO EXORBITANT FINE, NEITHER OF the McLaren drivers won the title that season. Alonso and Hamilton both finished one point behind Kimi Räikkönen. The only consolation was that they wouldn't have to be teammates for a single day longer.

But that didn't put an end to the hostilities. In 2008, Hamilton was openly recognized as McLaren's No. 1 driver, while Alonso was installed at Renault. The problem for Alonso was that in getting rid of his irritating teammate, he'd also traded in his competitive car. By the fifteenth Grand Prix of the year, in Singapore, Hamilton was back in the hunt for a first world championship title and Alonso hadn't been on a single podium all season.

It didn't look like the situation was about to change when Alonso started the race on a hot, muggy night at the Marina Bay Circuit in fifteenth place. An early pit stop only made things worse, sending him to the back of the field. Three laps later, however, Alonso's new teammate, the Brazilian Nelson Piquet Jr., appeared to lose control of his car and crashed into a wall, requiring the Safety Car to come out and slowly guide the pack around the track while marshals cleared the wreckage. This procedure happens all the time in a Grand Prix and is usually a useful moment to make a pit stop.

What made this particular Safety Car event so notable was that Alonso was now the only driver who didn't need a visit to the pits. So while everyone else stopped for fresh tires during those few laps under the Safety Car, he vaulted himself right back up the field, ahead of all the favorites. No one had a worse time of it than Ferrari's Brazilian driver Felipe Massa, who had been leading the race and was locked in a world championship dogfight with Hamilton. He accidentally drove out of the pits with the fuel hose still attached and found himself in last place.

Alonso went on to win the Grand Prix. But something about the way he'd leapfrogged his rivals smelled fishy to the FIA—and to Massa, who limped home in thirteenth position. The timing of the Safety Car just seemed a little too convenient.

By that evening, it was the talk of the paddock. Other team principals observed Alonso on the podium and found his celebrations strangely muted.

"That didn't look like a guy who, completely out of the blue, just won a race for the first time," says one former team principal, who watched from the pit wall. "He looked like a guy who had something on his mind. You know, Al Capone used to send hookers to his blokes and if they didn't get their leg over, he had them killed. I think Alonso would have struggled that evening."

In the end, it was Alonso's teammate Nelson Piquet Jr. who wound up getting whacked. Midway through the 2009 season, Renault boss Flavio Briatore made the call to dump Piquet after he failed to score a single point through ten races. Shortly thereafter, a curious story appeared in Brazilian media.

Although the FIA had never taken any action over the strange Singapore incident the previous year, Globo TV made the explosive claim that Piquet's crash in that race might not have been an accident after all. He'd been ordered to drive his car deliberately into a concrete wall at high speed to help Alonso improve his track position, setting off a chain of events that had potentially cost Massa a world championship and Brazil its first title since Senna.

For the third time in two years, Max Mosley called an emergency meeting of the World Motor Sport Council to investigate yet another shameful episode of cheating. If F1 was starting to look like a lawless, chaotic sport populated by ruthless maniacs with no ethical code or qualms about risking their drivers' lives, it's because it kind of was. Formula 1 seemed to be lurching from crisis to crisis. And the permanent quest for an edge, which had been the very essence of the sport, was morphing into something far more poisonous. The headlines were no longer about teams' innovating at the bleeding edge or the courageous feats of the men behind the wheel. They were asking whether F1 was broken beyond repair.

The FIA probe into Singapore soon showed that Renault had indeed asked Piquet to crash his car on purpose. Before the race, the team bosses had even sat down with him to point out on a map the precise corner where he should do it. Turn 17 was ideal, they told him, because there were no cranes there, meaning that any crashed car would take longer to clear away and require the deployment of the Safety Car. The plan was oddly brilliant, except for the fact that anything a driver did to crash on purpose would necessarily turn up in the car's telemetry readings. Specifically, the data showed that when Piquet entered turn 17 and felt his wheels spin, he had the most unnatural reaction for any driver who might have thought he was losing control of a car: he floored the gas pedal.

The council didn't have to deliberate long to decide that Renault was guilty. Denouncing the team's actions as being "of unparalleled severity," the FIA ruled that Renault had "not only compromised the integrity of the sport but also endangered the lives of spectators, officials, other competitors and Nelson Piquet Jr. himself."

There was no $100 million fine this time, however. Renault was handed a suspended sentence of two years, with the promise that another incident would lead to immediate disqualification. The man who really paid the price was Briatore: the former Benetton commercial whiz who'd overseen Michael Schumacher's earliest triumphs and Alonso's two world titles was given a lifetime ban from the sport (later cut to roughly three years by a French court). But even with Briatore gone—back to owning the Queen's Park Rangers soccer team alongside Bernie Ecclestone—the broader damage to the sport was indelible.

"What is happening here with Renault is more than just the tip of the iceberg, it is symptomatic of a wider problem," the former rally driver Ari Vatanen, who was running to succeed Mosley as FIA president, told Britain's *Sunday Telegraph*. "The image of the sport has been battered recently. Look at all the leaked dossiers. We have gone from 'Spygate' to 'Crashgate' with many other things in between. What the public see is a corrupt sport. They do not trust it and who can blame them."

Teams' willingness to go to any lengths to win had spiraled completely out of control. Between industrial espionage, intentional crashing, sprawling cover-ups, and a culture of total impunity, F1 had taken on a sinister patina at the worst possible moment. On top of the challenges that CVC was facing to keep the whole enterprise afloat in the midst of a global financial crisis, the terrifying collapse of the auto industry, the exit of tobacco money, and cratering sponsorship and broadcast revenue, the competitive legitimacy of the sport was now under threat as well.

"There is something fundamentally rotten and wrong at the heart of Formula 1," three-time world champion Jackie Stewart said. "Never in my experience has Formula 1 been in such a mood of self-destruction."

10

The Piranha Club

THERE WAS NO HIDING IT anymore: Formula 1 had become the most petty, most ridiculous, most vicious sport on the planet. The whole thing had nearly collapsed under the weight of two cheating scandals, orchestrated by egomaniacs, with so many outlandish details that it was hard to take it seriously as a global business worth backing, watching, or even rescuing.

What these episodes had shown was that F1 had become so divorced from reality that its central principle—a constant push and pull with the rulebook—now presented an existential threat. Spygate and Crashgate were beyond the pale. The ethos of bending the rules that had made F1 great had now broken it. There was cheating and then there was *cheating*. These were very much the latter.

And yet, as spectacular as these cheating crises were, there was a deeper, more systemic fissure in the sport that people were just beginning to grasp. It wasn't as sexy or as shocking as industrial espionage and intentional crashes. But this one posed an even greater risk to the future of the sport. Formula 1 was becoming financially unsustainable.

Not to its owners, of course. CVC was optimizing the business by creating order and process where Bernie Ecclestone had deliberately kept his operations as vague as possible. But for the teams themselves, the return on shipping a pair of cars and a few hundred employees around the world for most of the year, only to lose constantly, was fast approaching zero. The late 2000s weren't a great time to be in a sport that not only burned money to stay alive, but also hinged on constantly hitting up sponsors for more. Investors came and went so often that it's surprising certain teams could afford the paint to keep changing the colors. Any outfit that wasn't named Ferrari or McLaren could reasonably ask the question, What are we doing here anymore?

Take the team formerly known as Jordan. In the space of four years from 2005 to 2008, it successively raced under the names Jordan, Midland, Spyker, and Force India, in yellow, red, orange, and white. For those of you keeping score at home, that list of owners goes: an Irish motorsport enthusiast, a Russian steel tycoon, a maker of Dutch supercars, and an Indian alcohol magnate. The only thing that didn't change for the team were the results. No matter what name was on the car, the on-track irrelevance remained. Its drivers managed precisely one podium finish in four seasons.

The larger auto industry wasn't faring much better. In fact, it was in total free fall all over the world. In 2008 alone, the American giant General Motors saw its sales plummet by 41 percent. Honda sales crashed by 31.6 percent. At a time when no one was buying road cars, spending resources to race two F1 cars seemed even more irresponsible than usual. It wasn't much of a surprise at the end of 2008 when Honda announced that it was pulling out of Formula 1 altogether.

For those who stuck around, it began to look as if staying in the sport was now just a fast way to go broke. And no one knew this better than Williams, a team whose entire existence in F1 seemed to be spent on the verge of bankruptcy. They had managed to scrape by, of course, and even have a spell of wild success when Frank Williams allied two of the greatest loopholes in Formula 1 history in the 1990s: loading up the inside of his car with computers that made it nearly self-driving, and having the outside of the car designed by Adrian Newey.

But more than a decade after the heights of Alain Prost, Ayrton Senna, and the FW14B, technical loopholes weren't going to cut it anymore. In the era of astronomical costs, Frank Williams needed a different kind of egghead to unearth competitive advantages, not at the drafting table or in the wind tunnel, but in the team's balance sheet. Right around the time that a crummy baseball team in Oakland was turning around its fortunes on the cheap, Williams was hoping to Moneyball Formula 1.

The man he found to help him was an Eton- and Cambridge-educated logistics specialist from the mining giant Rio Tinto. Adam Parr was the man you called if you required insight on iron ore extraction in northwestern Australia. But if you needed someone to run your F1 team, he might not necessarily be your first pick. He knew next to nothing about motor racing.

The two men had first corresponded in 2000, when Parr wrote Williams an unsolicited letter to ask about how F1 teams went about moving around the world. They eventually met for tea in Perth, and soon Williams invited him out to a Grand Prix. Frank was struck above all by Parr's business sense—something that Williams himself badly lacked. He rang Parr late one night in 2002 to tell him just how impressed he was, and that he'd cooked up an idea.

"Adam, it's Frank," he said. "I just want you to think about one day running the team for me."

That day came four years later. Williams was sixty-four and neither he nor the team were in the rudest of health. Senna's death along with a purge of electronic driver aids had robbed Williams of its two greatest advantages. Though there were still world championships in 1996 and 1997, the real trouble started ahead of 1998. Williams no longer ran an Adrian Newey–designed car, since he had jumped to McLaren the previous year. And the team didn't have its preferred engine because Renault had quit Formula 1. As a newly public company, the French manufacturer simply couldn't justify to shareholders the absurd expense of being in F1 anymore—at least not for a couple of years.

That left Williams with Mecachrome engines, which were just old relabeled Renault engines. The following year, in 1999, they went to Supertec engines, which were just relabeled Mecachrome. From 2000 to 2005, Williams then partnered with BMW, which at least brought the team some victories again, but it remained a far cry from the heights of the mid-1990s. By the time Parr was considering Frank's offer, he was looking at an eighth-place outfit that had just posted one of its lowest ever points totals. It was also carrying some £30 million of debt. In less than a decade, the team had slid from the most technically advanced team on track to money-losing pack fodder. Against his better judgment, Parr took the job.

"I did no due diligence," he says. "It was probably a huge mistake."

Before he knew that, Parr was on a flight to Monaco that November for his first team principals' meeting. The man whose primary concern had so recently been the mineral composition of the Australian continent was now surrounded by a club of schemers who raced cars for money. Looking around at Flavio Briatore, Ron Dennis, and Jean Todt, Parr couldn't help but wonder what he was doing at the same table. These men hadn't just dominated the sport he was stepping into—they had shaped it. And they'd picked up at least a dozen world titles between them along the way. Still, Parr was certain that he possessed something they didn't: he came with fresh eyes and a fine understanding of how the real world did business. He wasn't like them—he didn't live to race—which he took to be a good thing. Parr thought Formula 1 could do with being a little less *Formula 1*.

Despite sitting trackside every other weekend and wearing a giant radio headset at work, Parr tried to go about his job like he would in any other corporate gig. He studied balance sheets, streamlined processes, and tried to identify new sources of income. Under the CVC ownership of the sport, teams were now getting a 50 percent share of the TV money, which at least gave them a larger piece of the pie. But running an F1 outfit remained one of the most efficient ways known to man to turn a large fortune into a small one.

Parr cast the widest possible net to capture new sponsors, just like his boss had done decades before. Frank Williams had courted Saudi princes in the 1970s. Now in the 2000s, Parr was spending weeks in Angola tying up a Luanda-based investment company called Ridge Solutions.

The other names on the car, each painstakingly convinced by Parr that the exposure on a losing F1 machine was worth millions of dollars, made up an eclectic mix. At the more traditional end were the Royal Bank of Scotland (which had helped CVC come up with the money to buy F1), the Brazilian oil giant Petrobras, and the Swiss watch company Oris. These were the kinds of sectors Formula 1 fans were used to seeing—money, energy, and luxury. A less familiar addition was the Hamley's toy store based in London, which splashed the logo of its website on the sidepods. But their check cleared and Parr wasn't in a position to turn people away. Signing any on-car sponsor amid global belt tightening counted as a coup. That's why Williams was more than happy to lend some visibility to a Hungarian energy drink that reeked of Budapest nightclubs. It was called Hell.

As for Parr's personal visibility, there was progress on that front too—namely, inside the Williams factory in Oxfordshire. Like any good manager, he wanted his staff to know he was right there in the trenches with them. Parr started parking his car every morning at the farthest end of the Williams lot from his office. He would walk in through the entrance by the wind tunnel, visit with the aero department, and then come through the various engineering sections to show his face and engage with the day-to-day running of a motor racing stable. Parr even moved his desk from near Frank Williams to a dingier office that overlooked the machine shop.

The move helped put him closer to the people he employed, even if he didn't have much material advice for anyone building a car. "Everything I suggested would have been regarded as bollocks, really," he says. But it gave him a close-up view of a team that simply didn't have the resources to keep up. The decisions that were made at the development stage were the choices Williams needed to live with all season. Upgrades were limited. And overhauling a chassis mid-campaign at massive expense—a lever McLaren or Ferrari could pull if they needed to—was never on the table.

"You have to be kind of super honest and realistic about where you are," Parr says. "But you have to also have hope and a belief that there are things you can do."

Parr wasn't an engineer, so he couldn't help there. Nor had he been a driver, so he had no on-track insight either. But he *was* a lawyer by training. If there was one area with room for major improvement at Williams, it was on the political front. "Frank had never challenged another team," Parr says. "And he'd never challenged the FIA on any decision. When I got there, we changed that."

Parr took a punchier approach than his aging boss. He'd struck up a friendship with Ferrari's Ross Brawn, who opened his eyes to the dark arts of the FIA appeal. If another team had found a loophole you hadn't and it was making them faster, then it was worth protesting. And if you caught the FIA in the right mood, you could at least ruin a rival's day. But as Parr familiarized himself with the internecine off-track battles of Formula 1, he realized that there was a bigger fight brewing in all of this. No matter how many protests anyone lodged or how many new engineers a team hired, changing the pecking order of this sport seemed impossible. Ferrari stayed Ferrari. McLaren stayed McLaren. No technical loophole could close the gap created by a $200 million budget. The competition itself was degrading, because two giants were proving impossible to catch through the methods that had defined the sport for half a century. Everyone else was going broke just trying to float in their orbit.

"We had moments where things were bearable on the track, but we had a lot of tough, tough, tough race weekends," says Parr, who attended around eighty Grands Prix in five years and watched his drivers finish on the podium just twice. "Financially, it was tough, because of not being able to give the engineers everything they wanted, or even anything comparable to the other teams."

Williams wasn't alone in this struggle.

For the teams and manufacturers still gambling on F1, reinventing the way they did business became a matter of survival. Parr shifted to spending most of his time on equalizing the playing field. He knew he could never bring his team up to the level of the large manufacturers with a budget that was barely a third of the size. What the sport needed was a new middle ground to make sure that the big boys spent a little less, the races could be more competitive, and the whole sport could be financially viable.

The trouble was that so much of the business came down not to raw numbers but to the relationships between three key factions: Bernie Ecclestone, who controlled F1's commercial rights; Max Mosley, the sport's regulator; and the team principals. Collectively, they were known as the Piranha Club. These weren't exactly people who worked according to best practices taught at Harvard Business School, or even normal human reasoning. During one tug-of-war over the terms of a new Concorde Agreement, the contract that binds the teams to Formula 1, Ecclestone bet Ron Dennis $100,000 that the deal would be signed by the end of that year. Dennis scoffed—he had major reservations about practically every page. This bet was easy money.

The moment Dennis took it, Ecclestone grabbed the papers, autographed the last page, and informed Dennis that he now owed him a hundred grand. Dennis stared back. "I never said you'd sign it," Bernie said, grinning.

Parr watched the scene play out and wondered to himself how anyone could get anything done in this mess. Still, he was prepared to try. If there was a glimmer of hope for saving all those F1 teams who were spending themselves into black holes for the privilege of losing, he was prepared to chase it. And after a couple of years in the sport, he at least had a sense of a solution.

Formula 1 needed a budget cap.

LIKE ANY CONVERSATION ABOUT THE SPORT'S RULES, THE DIScussion of limiting how much teams could spend came with a fight. Only this one threatened to blow up the whole sport.

Adam Parr hadn't been the only one to realize that Formula 1 was driving itself over a cliff. In 2008, shortly after Spygate, Max Mosley addressed a letter to the ten teams laying out a simple fact that they all knew to be true, whether they admitted it or not.

"Formula 1 is becoming unsustainable," he wrote. "The major manufacturers are currently employing up to 1,000 people to put two cars on the grid. This is clearly unacceptable at a time when all these companies are facing difficult market conditions."

Put in such stark terms, the whole enterprise felt a little bit absurd. Mosley went on: he demanded a 50 percent reduction in costs by the manufacturer teams (meaning those actually producing engines, like Ferrari, as opposed to those buying them from others). Independent teams, meanwhile, needed to become financially viable over the long term. There was no longer room for loss-making vanity projects that went up in smoke the moment their owners' interest faded.

At a time when every company on the planet seemed to be telling its staffers to do more with less, the FIA was asking the same of its Formula 1 teams. It wasn't enough to do things more cheaply. The sport also needed to repair its image as a wasteful pastime for idle millionaires and become "less profligate in its use of fuel." The challenge Mosley set out was to cut consumption in half by 2015 while maintaining the same speeds on track. Then some of this technology could find its way into more fuel-efficient road cars. F1 needed to remember its purpose as the automotive world's preeminent R&D department.

All of this, by the way, needed to be achieved, Mosley told them, "without affecting the spectacle in any way . . . The matter is now urgent."

Exactly how to pull any of this off would be left up to the teams, which could submit proposals for new rules that would come into force in 2010. If this turned out to be a task that was beyond the teams, then the FIA said it would be happy to lay out those rules for them.

In the meantime, it helpfully suggested some areas where these bloated organizations might start. They could set caps on time spent using computer-based analysis of aerodynamic systems. Or they could wean themselves off their addiction to wind tunnels (*buona notte*, Ferrari power station). Teams might also consider making more parts standard across the sport, like gearboxes. And if they knew what was good for them, the manufacturer teams would supply cheaper engines to client teams at a cost of around 2 million euros.

Gentlemen, start your calculators.

The ten teams' first course of action was to scramble together and do what they had done better than any group in any sport for half a century: they created an acronym. This time, it was FOTA, the Formula One Teams Association. Their first chairman was none other than Ferrari's own Luca di Montezemolo. And perhaps for the first time in Formula 1 history, every single team appeared to be united in a common cause.

Shortly after the 2008 Italian Grand Prix, they got to work. And shortly after that, the group began to fray.

Though everyone agreed that a cost cap was a good idea, each team seemed to have its own view on what shape it might take. For Parr and anyone else who stared at balance sheets hoping for miracles, a lower threshold made a lot of sense. But Montezemolo, the aristocrat, felt that any one-size-fits-all approach to spending limits would merely hurt the sport's most popular teams and damage the competition. He proposed that each outfit cut its budgets for 2010 by chopping their 2008 budgets in half. That would still allow Ferrari to blow over $100 million a year on its car, but also save smaller outfits from spending themselves into the ground. Montezemolo even suggested that the works, or factory, teams could build a third car for a less well-off partner if they had to.

By then, however, Mosley had convinced himself that the measures needed to cut much closer to the bone.

His ultimate goal was to encourage new investors to come in and expand the field of F1 teams to thirteen from ten. So the FIA's opening offer was to cap budgets at £30 million (around $42 million). For those who wondered how such a draconian slash would be possible, Mosley noted that this wasn't a hard cap—anyone who wanted to spend more would be allowed to. They just had to be prepared for the consequences: a series of restrictions designed explicitly to make their cars slower. These included fixed front and rear wings, along with a rev limiter on the engine.

Montezemolo felt that the FIA might as well have been forcing it to run its cars on three wheels. The situation, he said, was "grave and absurd."

Those kinds of numbers would require cutting three-quarters of the Maranello workforce. Worse still, they would cripple Ferrari, the only team that had participated in every single season of Formula 1.

The battle lines were drawn. The autumn came and went without the teams agreeing to new rules. And by spring, the fight had turned into a tug-of-war between Ferrari and the FIA for the soul of Formula 1—which suited both sides just fine, since both felt without the slightest doubt or humility that they were the only true stewards of the sport.

In the meantime, the US stock market had collapsed and the global economy was skidding fast into the guardrails. Mosley insisted that if they didn't take drastic measures soon, Formula 1 would simply cease to exist.

"We are confident (as are our accountants and lawyers) that a budget cap will be enforceable," Mosley wrote to the Scuderia. "The cleverest team will win and we would eliminate the need for depressing restrictions on technology."

Mosley wasn't shy about pressing his case in the papers. In an interview with the *Financial Times,* he even hinted at a future without Ferrari. "It would be very sad to lose them," he said. "They've been in the sport since the start, but if it's a choice between that and a situation doomed to failure and which would collapse F1 . . . we are not going to bend over backwards to keep them."

Now it was Montezemolo's turn to escalate. This was not the first time Ferrari had threatened Formula 1 with the nuclear option. In fact, they threatened it so often that it was widely understood to be little more than a bargaining chip. The possibility had come up most recently in 2005, during talks over the Concorde

Agreement, when Ferrari led the teams in asking for a larger piece of the sport's revenues and a greater say in how F1 was organized. That time, they had pulled back from the brink. What the other teams were unaware of then was that Ferrari had extracted a potent and baffling concession: a secret veto over any major future rule change. The Scuderia's status as the oldest, most powerful team in the sport had been privately enshrined into law.

Montezemolo couldn't fathom a series where the most successful teams would have their cars reined in. This was heresy, pure and simple—the motor racing equivalent of serving pinot grigio with bolognese. What would be the point of starting Grands Prix at such an obvious and artificial disadvantage? If that was going to be the case, they might as well do laps of the circuit with the handbrake on. Piero Ferrari, the son of Enzo, compared the proposal to allowing poorer soccer teams to put more players on the field. Whatever the metaphor, this couldn't stand.

Less than two weeks after Mosley's letter, Ferrari threatened to quit Formula 1 entirely. On the team's website, the Scuderia reminded the world what the sport would be losing with a piece headlined "The Pride of Making F1 Great." Ferrari was merely underscoring the lesson that Bernie Ecclestone had understood since his earliest meetings with Enzo himself. No racing series could ever call itself the pinnacle of motorsport if it didn't have Ferrari. The Prancing Horse wasn't just an occasionally successful team with a colossal sense of history and a huge following. It also carried a large chunk of the sport's legitimacy. Everyone knew that Ferrari was essential. Everyone except Max Mosley.

The FIA president wasn't impressed by the men from Maranello, nor was he prepared to budge. It seemed that no one was even discussing the details of the cost cap proposals anymore. In his most pointed statement of the crisis, Mosley took a direct shot at Ferrari's tactics.

"Good governance," the FIA said, "does not mean Ferrari should govern."

Even Montezemolo admits today that "personal differences" had played a role on both sides. The time for talking, press releases, and saber rattling in the newspapers was over. A last-ditch meeting between Mosley, Ecclestone, and FOTA's rebel alliance, held on board Flavio Briatore's superyacht, *Force Blue*, had amounted to nothing.

The teams braced for the worst. Renault warned its F1 suppliers that business might hit a speed bump if it decided to quit the series. Renault CEO Carlos Ghosn, years before he escaped an embezzlement scandal by fleeing Japan in a roadie case, announced that he was fed up with the whole imbroglio.

"Today we pay to be in Formula 1 and that is not normal," he said. "Intermediaries have made enough money. We want to take back control of Formula 1."

The teams delivered on that promise on June 19, 2009. Eight outfits led by Ferrari announced that they wouldn't participate in the 2010 season. After sixty years on the road, Formula 1 was heading for total implosion.

The only major team to stick with Mosley and the FIA was the one that could barely afford to go racing before the crisis: Adam Parr's Williams. In theory, they would be joined by Force India and three new teams whose entry into F1 was only made possible by the cost cap. But that ragtag collection of five could never pass itself off as any kind of prestigious motor racing series.

Not that anyone was convinced the FOTA operation would fare much better. Bernie, who was caught between supporting his ally Max and wanting his cash cow to stay alive, didn't think much of Montezemolo and Co.'s strategy. (Never mind that Bernie had built his entire base of power on a similar alliance of teams back in the 1970s.)

"They can't even run their own teams," Ecclestone said. "If the teams owned it, they would destroy it. They can't agree on anything, it would be a disaster."

He was right about one thing. Setting up a new series— basically F1, but without Bernie or the FIA—would require a Herculean effort. The FOTA teams would have to approach circuits, sponsors, and broadcasters and convince them to flip to an unproven product, merely on the strength of the teams' reputations.

They had no track record to sell, but Flavio Briatore was prepared to take on the challenge. He positioned himself as the new F1's very own Ecclestone. And in the space of a few weeks, he and Montezemolo planned to chip away at the empire that Bernie and Max had spent lifetimes building.

Decades of work would need to be crunched into a matter of months. And months of preparation for the following season would have to come together in a matter of weeks.

As it turned out, the FOTA project lasted only five days. While Formula 1 fans braced for the kind of split that nearly killed boxing, Montezemolo had kept the conversation going with the FIA's cornermen. During a final showdown in Paris, with Montezemolo, Ecclestone, and Mosley around a table, they reached a solution.

"Bernie understood that we were not playing a game," Montezemolo says. "We arrived one centimeter from the separation."

The FOTA teams promised to race in Formula 1 in 2010 and agreed to work together to control costs without a hard cap. The goal was to approach budgets of around $50 million within two years. The teams would also commit to staying in the sport until 2012 and fully recognize the FIA as the sole regulator. F1 could now reverse away from the abyss.

In exchange, at the end of a bitter fight that had turned oddly personal, Montezemolo got what he came for: Max Mosley agreed to step down as FIA president when his term expired that October. In a triumphant press conference, Montezemolo announced, "There is no dictator."

FORMULA 1 HAD INTRODUCED A COST CAP, BUT THAT DIDN'T make it cheap. Teams like Williams were still spending fortunes to achieve next to no success on the track. In 2010, its highlights were a fourth-place finish at the European Grand Prix in Valencia and a fifth place at Silverstone, which wasn't exactly what it hoped more than $50 million worth of car would buy. And that was before Williams had paid its drivers a single cent.

Under the new rules, driver salaries were excluded from the cost cap, though it's not as if teams had a ton of money to spend on them anyway. They needed to be creative: they would ask drivers, in effect, to pay themselves.

The way it worked was that teams would reserve one, if not both, of their seats for drivers who were skilled enough to handle F1 cars and appealing enough that they came with their own pre-existing sponsorship deals.

The concept had been around Formula 1 for decades, but never on this scale. Niki Lauda had borrowed money against his own life insurance to pay his way through his early seasons in the sport before being scooped up by Ferrari. Philip Morris had been paying the salaries of select drivers since the days of the Marlboro World Championship Team. And Frank Williams was known as one of the most prolific employers of pay drivers around: in 1975, he turned the Williams cockpit into a cash-spinning merry-go-round for his team, with eleven different men starting at least one Grand Prix that season.

By 2012, relying on pay drivers was a financial imperative for much of the grid. A full third of the twenty-four-man Formula 1 field consisted of drivers basically renting their seats. Sergio Pérez entered the sport in 2011 backed by a Mexican phone company founded by the country's richest man, Carlos Slim. Russia's Vitaly Petrov was believed to bring in around $5 million per season when he was hired by the Caterham team, which then signed a sponsorship deal with a Russian petrochemical firm. And French driver Romain Grosjean signed for Lotus with a promise to pay some $5 million provided by the energy giant Total.

At Williams that season, the situation called for both seats to be filled by pay drivers. In one car, they installed Bruno Senna, the nephew of Ayrton, because he came with an estimated $18 million worth of sponsorship, largely from Brazil. In the other car was the driver that Adam Parr spent the better part of two years chasing. His name was Pastor Maldonado, and he was driving for Venezuelan soft power. The backer this time wasn't a bank or a retail outfit or even an Angolan entrepreneur. This was the state-run oil company, Petróleos de Venezuela, which had recently discovered that the country was sitting on the world's largest proven reserves of crude oil, ahead of Saudi Arabia, Iran, and Iraq.

Maldonado, already twenty-five, had been putting in solid performances in GP2, the second-tier series just below Formula 1. But to get him into a Williams for 2011, Parr needed to agree to the terms with a Venezuelan government minister. Parr flew from London to Miami to Caracas to push the deal over the line. Except once he arrived, the minister stood him up. Only after Parr returned to the UK was he summoned back: "Now you come and meet me." So Parr did as he was told—London Heathrow to Miami International to Simón Bolívar International Airport, for the second time in a week. When he flew home this time, Parr had every sign-off he needed.

Maldonado was ecstatic to have finally arrived in Formula 1. What he'd actually signed up for, though, was slightly less exciting than expected.

"A lot of pressure for me from everywhere. From the team side, from the sponsors, from the country, from the supporters. Big pressure," Maldonado says now. "The biggest difficulty was that I was racing in different [circumstances]. And that was shocking to me. I had to adapt to these disadvantages.

"Our competitors, they had many more upgrades than us during the season, better packages, spending more on the technology side to develop the car during the season."

A major restructuring of the technical department didn't produce immediate results. Though the financial outlook of Williams improved following the team's stock market float in Frankfurt and the growth of the company's Williams Hybrid Power business, performance remained woefully out of step. In 2011, Williams finished ninth out of twelve teams, with 5 points, 645 behind the team that took the constructors' title.

Forces beyond Parr's control were beginning to move against him from inside and outside Williams. With negotiations on the next Concorde Agreement looming, Parr had run afoul of Bernie Ecclestone as one of the few team principals to criticize the Supremo in public. Bernie was now taking potshots right back. After the reshuffle at Williams, Ecclestone suggested that the real changes needed to come at the top. In fact, he had someone in mind to take on a larger role at the team and push Parr aside, a minority shareholder from Austria named Toto Wolff.

Ecclestone let it be known among the team principals—and specifically to Frank Williams—that the Concorde would not be signed as long as Parr was in charge. And this wasn't just Bernie double-talk to win a $100,000 bet. Parr found himself backed into a corner. Never mind that Williams had promoted him to chairman in 2010 and publicly called him his "natural successor" just weeks before. Now sixty-nine years old, Williams was no longer powerful enough to butt heads with Bernie. Parr couldn't be saved.

Parr had come into F1 thinking that a fresh set of eyes and a nose for efficiencies would provide the breakthrough that Williams needed. Maybe a couple of decades earlier, it would have been. But by the early 2010s, F1 required more than just a new outlook to win races, let alone championships. The press release about Parr's exit from Williams dropped that March, on the Monday morning after the second Grand Prix of the season in Malaysia. After five years of chewing, Formula 1 was finally spitting him out. Parr wasn't even around to see Maldonado win the Spanish Grand Prix that season, marking the team's first race victory in eight years. (This being Williams, the garage promptly caught fire during the celebrations.)

That single win was the moment Parr had been building toward during his entire tenure in the sport. And, typical of the F1 world he discovered, he'd been fired two months before he could pop a champagne cork.

"My wife always points out, well, what do you expect?" he says, looking back. "'Did you think you were going to work for the NHS or something?' It's called the Piranha Club for a reason."

IN THE CHAOS AND DRAMA OF FORMULA 1'S NOT-QUITE-breakaway, one man was still able to exploit a loophole. That was Ross Brawn's specialty.

If someone like Parr was an outsider looking to shake up the sport and save it with Moneyball tactics, only to leave jaded and exhausted, Brawn was the ultimate F1 insider. He knew engineering, he played politics, and he bent the sport to his will—all to become richer than his wildest dreams in the process.

The craziest year of Brawn's professional life began in late 2008 at the Honda Racing team. The Japanese manufacturer had been back in Formula 1 since 2006 and hired him the following year, fresh from his all-conquering stint at Ferrari, to reorganize their flailing operation. But the combination of the economic crisis and continually poor results cut short Honda's patience for F1. The company was ready to pull out.

Honda's decision to quit the sport hadn't entirely blindsided Brawn. He'd noticed over preceding weeks that Honda seemed to be repatriating key Japanese staff members from Brackley in England back to Tokyo. The confirmation of his fears came in November 2008, when Brawn and his business partner Nick Fry were summoned by Honda's bespectacled COO, Hiroshi Oshima, to a hotel near Heathrow Airport.

Oshima sat them down in a small, empty conference room and put his glasses on the table. He had news.

"It's not good—not good at all," he said. "I'm sorry, guys. We are stopping."

After three seasons of mediocrity and just one race victory, Tokyo was pulling the plug. Before that revelation could sink in, the man from Honda asked Brawn and Fry to follow him into a larger conference room where a team of lawyers and finance types were already sitting around a table, waiting to begin the executions.

Honda's instructions were to send everyone home and turn off the lights.

Brawn pointed out that things didn't quite work that way. They couldn't just shut up shop. There were employees and notice periods and, lest anyone forget, an F1 car in the works for 2009 that Brawn felt was just too good to throw away. In a last-ditch plea, he laid out an emergency plan that he and Fry had been working to cut their £200 million budget by 30 percent. But Oshima wasn't having it. The financial crisis was pummeling Honda and the company was expecting plenty of red ink for 2009, he said. US dealers were refusing to take new cars because they didn't see how they could sell them. Losses would be on the order of "three thousand million dollars."

Brawn went into problem-solving mode. What if Honda sold the team instead of shuttering it? And since the company was still on the hook for a little while anyway, would they continue paying the bills until a buyer emerged?

Remarkably, Honda said yes. That was Brawn's first miracle. The next one was striking a deal for Mercedes to replace Honda as the team's engine supplier. (The extent of Mercedes's F1 business in those days was selling power units to teams such as McLaren—it didn't yet have its own racing team.) Unlike Ferrari, which only offered Brawn the previous year's model, the Germans were happy to give him their most modern technology. The rub was that they demanded payment up front—8 million euros for the season. Brawn agreed to wire them the money in 24 hours.

The turmoil left Brawn's team in a strange spot. Not dead in the water, but not 100 percent alive either. The uncertainty set off a frenzy at the factory, where Zombie Honda mechanics and engineers pressed ahead to develop a car that might never see a single race. Yet they knew they didn't have much of a choice. If the team was going to be sold, then they needed to make sure there was enough *there* for someone to buy.

Over the weeks that followed Honda's announcement, plenty of suitors came out of the woodwork. Vijay Mallya, the Indian booze king, expressed interest in buying it. So did a former travel agent named Achilleas Kallakis—real name Stefan Kollakis—who was later convicted in 2012 of Britain's biggest ever mortgage fraud. Then came Richard Branson, the billionaire founder of the Virgin Group and sometimes balloon pilot. Brawn and Fry rolled out the red carpet for all of them at the Brackley headquarters. But none of these Formula 1 arrivistes felt right.

That's when another potential buyer emerged for this untitled Ross Brawn project: Ross Brawn himself. In early January, he and Fry approached Oshima with an audacious plan for a management buyout. The Honda people were skeptical. For most of the senior team in Tokyo, the only reason to be in F1 was to beat Toyota at all costs. And after posting one win in three years against zero for Toyota, they weren't exactly impressed with how things had been run.

Yet these were people that Honda knew. They were willing to take the entire F1 operation off their hands, liabilities and all, for £1. The majority shareholder would be Brawn, with just over half, and the rest would be split between Fry and other directors. In return, Honda would agree to give them enough cash to get through the 2009 season. Considering the team had no sponsors and no income beyond the £10 million in prize money they received from Ecclestone for the previous year, they needed a lot of it. Brawn and Fry told Honda they required a minimum of £100 million to get through it.

It's at this point that Bernie caught wind of what was happening. He didn't appreciate not being involved when a good deal was in the works and explored the possibility of swooping in to buy the whole team from under Brawn.

"Bernie did not like Ross," Fry wrote. "Bernie resented the fact that Ross was one of the few people in Formula 1 who wasn't beholden to him."

The struggle to woo Honda played out for over a month. Fry leaned on every contact he could, including the British ambassador in Tokyo. He persuaded him to deliver a letter directly to the CEO of Honda, begging him to preserve important jobs in Brackley by agreeing to the buyout.

The Brawn campaign proved convincing. On February 23, 2009, Honda's board rubber-stamped the sale of the F1 team to Brawn's group, barely over a month before the first Grand Prix of the season in Australia. The legalities dragged on until March 5.

And by now, the countdown to Melbourne had ticked down to three weeks. That time frame was just about manageable if everyone worked like crazy. The final preseason test in Barcelona in mid-March, however, was not. Brawn would have to skip the test entirely and go into the season cold.

That didn't do much to reassure anyone on the outside that the team was viable. Organizers in Melbourne were so dubious the team would make it to Australia that Brawn GP—a name Bernie personally objected to—was never listed in the race's official program. After all, no one had seen these cars. They weren't even certain that they existed.

The only people who believed in them were the people inside the factory at Brackley who'd been briefed on the secret weapon in the car's design. And once they saw the simulated lap times, they knew for certain what no one else could have imagined four months earlier. The Brawn GP cars existed—and they were fast as hell.

Their unbelievable combination of power and downforce went on full view from the first practice session on Friday in Australia. By Saturday's qualifying sessions, everyone understood that this wasn't a fluke. The Brawns took first and second place on the grid, which was exactly how they finished the race on Sunday too. Brawn GP became the first new team to occupy the top two spots on the podium in their debut race since Mercedes-Benz in 1954.

The rest of the F1 paddock quickly clocked what was happening. The Brawn secret weapon was a piece of aerodynamic innovation called a double diffuser.

Back in 2008, one of Honda's Japanese engineers had spotted a gray area in the rules governing the part of the 2009 cars where the floor comes to meet the rear axle. The precise functioning of it requires a few years of graduate-level physics to fully grasp. But the upshot is that the airflow through the double diffuser created a lower pressure zone under the car than a regular diffuser and sucked it closer to the ground. More grip equals more speed in the corners. More speed in the corners means more winning.

"It's a very clever device," Max Mosley, in his final months as FIA president, told reporters before the race, knowing this was about to become a problem for him. "You can make a good case for saying it's legal and a very good case for saying that it's illegal."

Two other teams, Williams and Toyota, had also stumbled across the double-diffuser interpretation of the rules and found similar leaps in performance. The Toyotas took third and fourth in Australia behind the Brawns, with one of the Williams cars finishing sixth. Ferrari and McLaren were nowhere. However briefly, the entire order of modern F1 had been flipped on its head.

"As you can imagine, it made us blink hard," writes Brawn's rival Adrian Newey, who immediately sought clarification from the FIA.

With that much downforce, he argued, these cars were surely a safety risk, weren't they? Yes, Newey was actually making the case that the Brawn GP car was dangerous and should be outlawed because it was *too fast*. Mosley assured him that he was right and the double diffusers would soon be ruled out. But the longer the examination went on, the less confident Newey became. A formal protest went nowhere. And that April, Brawn received the green light to keep dominating the field. The team's lead driver, a twenty-nine-year-old Brit named Jenson Button, won six of the first seven races.

"You have built me a monster of a car," Button crowed over the radio that spring in Turkey.

That Button, of all people, was leading the world championship standings was all the proof that Brawn's critics needed to claim that the car was a freak of engineering and ruining F1. Button was known as a nice guy, but he was no Ayrton Senna. This was a driver who had exactly one race win to his name in a decade of trying before Brawn built him that monster. His teammate, meanwhile, was the thirty-six-year-old Rubens Barrichello, whose best years in F1 had come as the No. 2 Ferrari driver behind Schumacher and who'd almost hung up his helmet before the season. While Brawn GP sprinted away, the trio of former world champions in the field (Fernando Alonso, Lewis Hamilton, and Kimi Räikkönen) had zero podiums to show for themselves in the first five races.

For a sport that had a real problem shaking its reputation for being decided by minor, arbitrary rule changes, this wasn't a great look. Critics howled that titles were being settled in FIA tribunals and closed-door meetings, not on the track.

"Our drivers are, or have been, world champions," an outraged Flavio Briatore told the Italian press shortly before being chucked out of the sport over Crashgate. "And then you have a driver who was almost retired [Barrichello], and another who is a concrete post [Button], fighting for the championship. I don't know how we can say we have credibility."

Luckily for the sport, the truth of any devastating F1 advantage is that it can't last long, especially given the financial disparities that now existed between the top teams and the rest. Back in the days of Williams's dominance of the Ferrari dynasty, uncovering a technical loophole conferred an advantage that lasted for a season or more. What Brawn was discovering now was that it barely bought you a couple of months. Money had changed the calculus. Brawn GP's gap on the field was already starting to close by the halfway point of the season. Some of its rivals were blowing £2 million a race on new parts to catch up, while Brawn didn't have a spare dime to spend on further development—or anything else for that matter. Unlike other team principals or owners, who flew around the world in Gulfstreams, Ross Brawn was making his way to European races on Easyjet.

That's why it wasn't until the penultimate Grand Prix of the season that the most dominant car anyone had seen in years finally clinched the title. Button was crowned in October, despite not winning a single race after June as nerves and a series of on-track mishaps slowed his progress. But he scraped together enough points to get over the line. Brawn GP, the team that was supposed to turn off the lights ten months earlier, now had drivers' and constructors' championships.

For Ross Brawn, the campaign was pure vindication of the riskiest bet of his life. It also came with one enormous question. *What now?*

The Honda money had run out. Development on a 2010 car was all but nonexistent. And Brawn had to figure out what to do with a team that was built only for a short stint. As he looked around for a solution, he found that the answer was already inside his car. He needed to speak to Mercedes.

The German automaker hadn't suffered from the global downturn quite as badly as some of its industry rivals. In 2009, Mercedes's parent company, Daimler, had been able to raise cash by bringing in a major shareholder from Abu Dhabi to stabilize the business. So when Brawn and Fry pitched Mercedes on buying their team, F1 no longer seemed like the stupidest investment it could make.

One look around the sport's landscape was enough to tell them that they'd have a chance to win almost immediately. They'd be buying the defending world champion team at a time when every other manufacturer seemed to be fleeing the sport. Honda was out, while Toyota and BMW announced that they were leaving the sport after 2009. Renault was frustrated too, and announced plans to scale back its involvement in F1 by selling off its factory team and becoming just an engine supplier. The smaller teams, meanwhile, were running off-the-rack Cosworth engines. As for Ferrari, the Scuderia famously couldn't get out of its own way.

The executives in Stuttgart could see that if Mercedes wanted to pick up championships again for the first time since the 1950s, there was no better moment to get back into Formula 1.

That November, Mercedes announced that it was dumping its partnership with McLaren and taking over 75 percent of Brawn GP. The price tag was around 200 million euros, suddenly making Brawn wealthier than he'd ever imagined. Plus, with Brawn agreeing to remain in charge, success seemed all but guaranteed. After half a century on the sidelines, Merc was back, ready to steamroll a field of motor racing also-rans who couldn't boast a fraction of their speed or heritage.

What Mercedes hadn't counted on was just how fast some cars built by a fizzy drinks company could be.

11

Running of the Bulls

DIETRICH MATESCHITZ NEVER INTENDED TO build a Formula 1 world champion. But then, he never intended to build a global energy drink empire either.

How he managed to pull off two such unlikely feats despite harboring no lifelong ambition to accomplish either one can be traced back to the same basic impulse: Mateschitz was feeling bored.

In the early 1980s, it was the professional grind of the personal hygiene game getting him down. As a marketing executive in his native Austria, Mateschitz traveled the world hawking toothpaste, detergent, and women's cosmetics. He was stuck in a rut, fast approaching forty and tired of life on the corporate treadmill. "All I could see was the same gray airplanes, the same gray suits, the same gray faces," he said later. "All the hotel bars looked the same, and so did the women in them."

All that gray might sound like a Ron Dennis fever dream, but Mateschitz was desperate for a way out of this monochrome nightmare. He found it on a business trip to Thailand in 1982. Searching for something to take the edge off his jet lag, Mateschitz was offered a local pick-me-up that was a favorite among long-haul truck drivers: a syrupy, yellowish concoction cooked up by a Thai pharmacist as a cure for hangovers. It hit Mateschitz like a triple espresso shot to the eyeball. "One glass and the jet lag was gone," he told *The Economist*. Within months, he had quit his job and set up a company to start selling this miraculous stuff in the West.

Before it reached the shelves, though, Mateschitz made a few tweaks to the recipe—the viscous truck-driver juice might have been a little too intense for the general public. Mateschitz wanted a carbonated, less concentrated version of the original, so he stripped the ingredients down to the essentials: an amino acid called taurine, the carbohydrate glucuronolactone, and an industrial dose of caffeine. Then he redesigned the traditional soda can into an eight-ounce bullet shape and priced it at $2 a pop to signal that this wasn't just another cola. He also opted for a slightly catchier version of the Thai name, Krating Daeng, which loosely translates to "Red Water Buffalo."

By 1998, Red Bull had single-handedly created a whole new category of beverage called "energy drinks," was selling three hundred million cans worldwide, and had made Mateschitz a billionaire.

The secret to Red Bull's success wasn't the flavor (bad) or its health benefits (worse) but the ingenious way it was marketed. Pitching it less as a sports drink and more as a symbol of an edgy, carpe diem, YOLO lifestyle, Mateschitz largely dispensed with traditional print and TV ads in favor of low-key grassroots campaigns focused on music festivals, wild parties, and, above all, extreme sports. He plastered the Red Bull logo on skateboarders, base jumpers, cliff divers, ice climbers, ultramarathoners, waterfall kayakers, and all over an oddball event in which thrill-seeking lunatics launched homemade flying machines off the end of a pier. All of which helped to finance Mateschitz's taste for acquiring his own aircraft—a historic collection that included a DC-6B once owned by the Yugoslav dictator Marshal Tito—and his personal desire to own a private island in Fiji. "Doesn't everybody want their own South Pacific island?" he said in an interview with Bloomberg.

Yet none of those endeavors embodied Red Bull's quintessential ingredients of speed, endurance, and the imminent threat of mortal danger quite like Formula 1. Two decades after Marlboro execs looked at F1 drivers and saw the heirs to the American cowboy, Mateschitz recognized them for what they really were: overcaffeinated adrenaline junkies with scant regard for their personal safety. It was a match made in marketing heaven. In 1989, the Austrian driver Gerhard Berger became the first professional athlete to be sponsored by Red Bull. Six years later, the company took over as title sponsor of the Swiss F1 team Sauber. And in 2001, Red Bull established a driver development program to identify young prospects with F1 potential. One of its earliest finds was a twelve-year-old karting gem named Sebastian Vettel.

In a little over a decade, from 1989 to 2001, Red Bull had transformed itself into part of the furniture of the Formula 1 paddock. Even other billionaires with a gift for marketing and a thirst for speed were surprised by how quickly Mateschitz ramped up his investment in the sport. "I said he's never going to come into Formula 1," Bernie Ecclestone remembers. "He got just as much publicity from guys jumping out of balloons and whatever. But he's very, very good at building a brand."

That brand was now inextricably linked to the world's premier motor racing series. Yet Mateschitz wasn't entirely satisfied—boredom came easily to him. For one thing, the Sauber team's struggles began to concern him as his drivers sank into mediocrity. Initially, the pairing of his upstart brand with a rank outsider had seemed like a natural fit. But after managing just five podium finishes in seven seasons, he wondered whether being associated with a perennial also-ran might not be so great for business. "If an insurance company sponsors a team, and the team loses, people don't change their insurance company," Mateschitz liked to say, "but when the Red Bulls lose, people get a new drink."

More than that, he wasn't thrilled about being one more corporate logo in the F1 pit lane. Red Bull's association with extreme sports was all about getting out there and making stuff happen. The company didn't just sponsor athletes, it dreamed up and organized most of the events they competed in too. But as Mateschitz looked at Formula 1 in the early 2000s, he felt as though Red Bull was becoming just another name on a spoiler. The whole sport, he said, had been reduced to "BMW against Mercedes and Honda against Toyota." It was starting to look an awful lot like all those gray airport lounges. Formula 1 needed a serious injection of taurine.

In 2005, at precisely the time when other edgy lifestyle brands with questionable side effects—Marlboro, Camel, Lucky Strike, and the rest—were being forced out of the sport, Mateschitz bought the husk of Ford's Jaguar F1 team for one pound sterling and renamed it Red Bull Racing.

Once again, his first instinct was to tweak the recipe. While the rest of the paddock obsessed over being the fastest team in F1, Red Bull happily focused on being the loudest.

"We always thought that if the old teams did something one way, is that necessarily the way Red Bull would do it?" says Dominik Mitsch, Red Bull's head of F1 marketing operations at the time. "And if it's not the way we would do it, why couldn't we change it?"

The team made that intention clear at the opening European race of the 2005 season, the San Marino Grand Prix, which marked the first opportunity for teams to bring their own motor homes to the track. Red Bull's was impossible to miss. Motor homes had received some significant upgrades since the days of Frank Williams operating out of his family caravan, but they were still little more than glorified portacabins, parked in the paddock for the express purpose of schmoozing sponsors and holding technical meetings. Red Bull had something a little different in mind. When the teams pitched up at Imola, they were confronted with a gleaming three-story glass-and-steel edifice with a hydraulic roof that lit up at night to reveal a nightclub on the upper deck. It was named the Energy Station, but in truth it looked more like a spaceship had landed in the paddock. As fate would have it, it had touched down right next to the McLaren motor home.

Ron Dennis almost had a seizure.

Once the race weekend got underway, it turned out that the most radical part of the Energy Station wasn't the design. It was the door policy. The team motor home was supposed to be one of the most sacred and private enclaves in the paddock, a place where deals were made and race strategies were devised. But Red Bull threw its doors wide open and welcomed everyone—mechanics, drivers, engineers, team principals. It didn't matter who you were or which team you worked for, anyone with a paddock pass around their neck was welcome to attend their "Chilled Thirstday" parties on the Energy Station roof, where guests could listen to ambient electronic dance music while knocking back a couple of beers or Vodka Red Bulls.

Naturally the other teams viewed this display of warmth and hospitality with deep suspicion. None more so than McLaren, whose employees were expressly forbidden from venturing inside the Energy Station right next door—not that the embargo stopped them. In classic Formula 1 style, the McLaren workers soon discovered a loophole. "They'd change their clothes and put on their own shirts to come in, rather than come in wearing a McLaren shirt," says David Coulthard, the veteran F1 driver who became one of the first new additions to Mateschitz's Red Bull team. "And in that way, it sort of opened up the paddock, which had been a very segmented place."

The Energy Station was just the start. Red Bull soon established itself as the sport's foremost disruptor, seemingly determined to challenge every unwritten rule and standard way of operating. No piece of conventional thinking came under greater attack than the notion that Formula 1 should be a deeply serious endeavor. Red Bull had the bonkers idea that F1 could actually be fun. So in one piece of cross-promotion, the pit crew dressed up as *Star Wars* storm troopers. The team also blared music out of its garage at all hours. It arrived at every race with a small army of young women known as "Formula Unas," chosen through a beauty pageant to act as brand influencers. And it hired a Fleet Street veteran named Norman Howell to produce *The Red Bulletin,* a muckraking tabloid whose editorial mission was "taking the piss out of the sport," Howell told the *New York Times,* "because the sport is very up itself."

As far as Red Bull was concerned, nowhere was more up itself than the opulent capital of Formula 1 excess known as Monaco. Which is why the team reserved its starkest break with tradition for an early visit to the principality. Aware that its new Energy Station wouldn't fit in the tight confines of the Monte Carlo paddock, Red Bull opted to relocate to somewhere a little more spacious: the Mediterranean Sea. The team commissioned a ginormous floating pontoon, which was assembled and stored in the Alps until the weeks before the Grand Prix, when it was hauled to Imperia in northern Italy. There, some seventy engineers spent twenty-one days constructing a new Energy Station on top of the pontoon—this one equipped with a DJ booth and a swimming pool—and a further six hours sailing the entire monolith forty miles down the coast, where it docked in the Monaco harbor, right next to the superyachts. When Mateschitz came aboard ahead of the race, the bigger, bolder, and substantially more buoyant Mk. 2 Energy Station was floating in place for its unveiling. Mateschitz swanned around, hobnobbing with George Lucas, and even found enough room for an impromptu kickabout with the Brazilian soccer star Roberto Carlos.

This was more like what Mateschitz had in mind when he invested in Formula 1. In fact, he was having so much fun that he soon decided that owning one F1 team wasn't enough—never mind that buying a single team had often been sufficient to cripple a billionaire's finances. Less than a year after establishing Red Bull Racing, Mateschitz acquired a second team, the outfit formerly known as Minardi, and renamed it Scuderia Toro Rosso, in part to give the graduates of Red Bull's driver development program a clear pathway into the sport. This was a team planning for the long term.

Red Bull had once been dismissed as a marketing gimmick, a noisy intruder that would quietly slip away after a couple of seasons, a few spectacular crashes, and several hundred million in expenses. But it was now clear that Mateschitz and his Red Bull empire weren't going anywhere—and they certainly weren't going anywhere quietly.

Mateschitz, meanwhile, had come to a realization of his own. The Energy Station and *Star Wars* tie-ins had been good ways of helping Red Bull stand out from the crowd. Yet even before he became a fully fledged F1 owner, he knew that only got you so far. To make some noise in the sport, there was only one marketing ploy that would really pay off. Red Bull Racing needed to win. To achieve that goal of transforming his brash outsider into a genuine contender capable of toppling F1 bluebloods like Ferrari and McLaren, Mateschitz had made what may have been the most audacious gamble of his career.

At the very outset of Red Bull Racing, he hired the youngest team principal in F1 history, handed him the keys to the entire operation, and got out of the way.

CHRISTIAN HORNER ROSE TO HIS FEET ONE AFTERNOON IN January 2005, cleared his throat, and looked out at the sea of faces staring back at him from the factory floor. It was at precisely that moment, as he scanned the crowded room, that it occurred to him for the only time in his life that maybe he wasn't cut out to lead a Formula 1 team after all.

Which was awkward, because only a few hours earlier, Dietrich Mateschitz had summarily fired the managing director and technical director of his newly acquired Red Bull Racing and placed Horner in sole charge. The thirty-one-year-old had never driven an F1 car, never worked for an F1 team, and possessed no college degree or expertise in engineering, design, or aerodynamics. But something about his keen understanding of how to marshal the resources of a racing outfit in the lower categories of motorsport had impressed Mateschitz. The rest of the Red Bull staff, however, would require more cajoling.

As Horner prepared to address the hundreds of complete strangers who made up his new workforce, he could sense that he wasn't the only one having some doubts. It had been an unsettling day for everyone and the emotions were written across the faces of each individual in the room. There was anger, there was confusion, there was anxiety. And now, as Horner stood before them as their new boss, there was also an unmistakable sense of skepticism. *Who the f— is this lad?*

In truth, Horner wasn't a complete unknown. In British motor racing circles, he'd been regarded as a promising driver in the junior series, graduating from go-karts to Formula Renault and ultimately rising all the way to Formula 3000, as the tier directly below F1 used to be known. It was there that Horner's ambition to fulfill his childhood dream by becoming a champion F1 driver had taken a turn. The problem was it had taken that turn more slowly than Horner would have liked. Coming out of the pit lane during a preseason test at the Estoril racetrack in Portugal in early 1998, Horner found himself right behind the Colombian driver Juan Pablo Montoya headed into Turn 1, a high-speed bend with almost no run-off.

He watched as Montoya threw his car into the corner flat-out, foot planted hard on the throttle. As he flew through the bend at 130 mph, Montoya was inches from the guardrail, with the car at an angle that seemed to defy the laws of physics. The lateral force was so great that it looked as though the outside rear tire was about to tear itself clean off the rim. Montoya didn't so much as flinch. Horner gasped inside his helmet.

"I knew then that I didn't have it," Horner says. "There was something between my foot and my brain, some kind of self-preservation, that wouldn't let me do it, wouldn't even allow me to try. I knew there and then that it would be my last season as a driver."

It wouldn't be his last season in motorsport, however, because Horner happened to be driving for his own team, Arden International, which he had established the year before with a loan from his father. Once Horner realized he wouldn't make it as a driver, he did exactly what any responsible team owner would do: he fired himself and found a replacement.

Fortunately, those replacements weren't weighed down by trivial concerns like self-preservation. Horner's Arden team swept to three consecutive Formula 3000 titles. And by 2004, the driver turned team principal felt he was ready to make the step up to Formula 1.

At least one influential figure in F1 felt the same way. Believing that F1 needed an injection of youth, Bernie Ecclestone encouraged Horner to explore a takeover of a team then owned by Eddie Jordan, an Irish entrepreneur who was looking to sell up. Horner spent time touring the team's facilities, looking through the books, and putting together a proposal. He submitted an offer to Jordan, pledging to take on all of the team's liabilities in exchange for a token price of £1. The response Horner received from Jordan was short and to the point. "He screamed at me for a few minutes," Horner says, "and I think most of the words started with F."

It looked as though Horner's F1 ambitions would remain on hold until he received a call in late November 2004, two weeks after Red Bull had completed its acquisition of the Jaguar team. The voice on the line belonged to Helmut Marko, an Austrian former race driver who now managed Red Bull's junior program. He explained that Mateschitz wanted to meet with Horner. A few days later, the two men sat down for lunch in Salzburg.

"I've got big ambitions with this team," Mateschitz told Horner. "I want it to be different, I want it to have a different energy. We're not going to be corporate, we're going to do things the Red Bull way."

The Red Bull way meant being unafraid of failure. It meant trusting your gut, giving youth an opportunity, and occasionally rolling the dice. "And I'm prepared to take a chance now," Mateschitz told him. "I'm willing to take a risk on you."

By the time the waiter brought the Linzer torte for dessert, Horner had already made up his mind. This was the opportunity he'd been waiting for. But just a few weeks later, at the general meeting inside the Red Bull factory, the scale of the undertaking was starting to sink in. Instead of the workforce of 25 he'd been responsible for at Arden, he was about to address a room of 450 people, most of whom had the distinct impression that Christian Horner was completely out of his depth.

He used his introductory speech to try to convince them otherwise. Horner expressed his excitement for the challenge ahead, his determination to make Red Bull a success, and his firm belief that his experience in Formula 3000 would help him to build a winner in Formula 1. "It's a big job," he told them. "But the basics remain the same. It's all about people—the people in this room are our biggest assets."

Once he was finished, Horner sat down. The people in the room promptly walked out in protest.

Horner returned to his office with the facility practically deserted. His predecessor's Christmas cards and a half-drunk cup of coffee sat on his desk. Outside, his secretary was in tears. There were eight weeks until the start of Red Bull's first season.

"Okay," Horner said to himself, "this is the stuff."

HORNER'S STAFF EVENTUALLY RETURNED TO WORK—THEY had kids and mortgages, after all. He began to evaluate what he needed to do to improve on the team's showing the previous year, when Jaguar finished seventh in the constructors' championship, with a grand total of 10 points. At first glance, it seemed like it wouldn't take much. In Red Bull's debut race, the 2005 Australian Grand Prix, David Coulthard finished fourth, No. 2 driver Christian Klien finished seventh, and the team had 7 points on the board. By the end of the second race, Red Bull had already eclipsed Jaguar's entire season total.

Horner, whose initial two-year contract at Red Bull paid him a bonus for every point scored above Jaguar's 2004 mark, was pulling out every stop to maximize the team's performance. He rotated the drivers of the No. 2 car and placed special attention on speeding up pit stops. The team spent so much time running its Cosworth engines in higher power modes than recommended by the supplier that the screams from Cosworth executives to turn the engines down became the running soundtrack to their season. "It was always about pushing the boundaries," Horner recalls, "and taking those irresponsible risks." By the end of the season, Red Bull was seventh again, but with 34 points.

Horner knew Red Bull was still a long way from being a contender. To build a team capable of challenging for wins and championships would take more than simply running the engines into the red for half the season. The team needed stability, it needed more engineering know-how, it needed more technical expertise, more skilled drivers, and more horsepower. Every time Horner looked over his checklist, he asked himself the same question: *Where on earth do I start?*

His first priority was greater technical leadership and direction—an area where Horner's lack of formal qualifications left him slightly ill-equipped. The technical operations director he'd inherited was a passionate Italian from the Alps adjoining Austria, who possessed a strange accent and a virtuosic command of expletives, but wasn't quite so knowledgeable when it came to aerodynamics. His name was Guenther Steiner. And as Red Bull's debut season progressed, Horner quickly came to the conclusion that Steiner needed to be replaced. "Guenther was—and is—a character," Horner says, "but it was obvious he was not a technical leader."

Instead, Horner set his sights on the most highly regarded aero expert in the paddock—who also happened to be his new neighbor. The fact that Red Bull's Energy Station was situated directly adjacent to the McLaren motor home had felt like the universe playing a cosmic joke on Ron Dennis, but the punch line was devastating: Horner would use his proximity to McLaren's headquarters to steal away Dennis's most prized employee.

Horner's first encounter with Adrian Newey came about entirely by chance. It was at the San Marino Grand Prix in 2005 and Newey had been curious to see what Red Bull's Energy Station was all about. That evening, he snuck over from McLaren, where he held the post of technical director, for a closer look and bumped into Horner, who showed him in for a drink. Their next meetings that season were far less coincidental.

They happened because Horner began actively stalking Newey in a bid to engineer as much facetime as possible. "His tactic was to build a relationship with me by 'accidentally' bumping into me in the paddock," Newey wrote. "I'd be walking one way and he'd happen to be going in the other direction. 'Oh, hello Adrian . . .'"

That May, Newey received an invitation to attend the premiere of *Star Wars: Episode III—Revenge of the Sith,* and found that he happened to be sitting right next to Horner. Weeks later, he also happened to find himself on the floating Energy Station in Monaco. And in July, he happened to find himself sitting down for dinner at the Bluebird restaurant on London's King's Road with Horner and Coulthard. The two of them pitched him hard on Red Bull's ambitions, though by now it was clear that Newey was at least entertaining the thought of leaving one of the sport's most storied teams for the F1 equivalent of a start-up. The only question that hadn't been discussed at that stage was money.

Those discussions were reserved for October, when Horner and Newey flew to Salzburg and then took a helicopter into the Alps for a casual lunch with Mateschitz. "We showed him something of what Red Bull is like," Horner says of the get-together. "Among other things, we stuck him in our Alpha Jet fighter and he went inverted over Kitzbühel at five hundred feet."

If Newey was feeling a little queasy after his ride, it wasn't half as bad as Mateschitz felt the following day when he discovered how much the world's highest-paid racing designer was expecting to take home. Unsure that Newey was worth the seven-figure price tag, Mateschitz made a call to Gerhard Berger, Red Bull's original Formula 1 driver and one of his closest confidantes.

"Gerhard, we have Adrian Newey here in Salzburg, but he is very expensive," Mateschitz explained. "What should we do?"

"Well, it depends," Berger replied, "on the value you put on a second a lap."

Adrian Newey was formally introduced as Red Bull's chief technical officer in November 2005 with the stated aim of building the team into a championship contender. Within seven races, the team achieved its first podium, when Coulthard came home in third at the 2006 Monaco Grand Prix. The celebrations that night were as gleefully excessive as you might expect. Some thirty years after McLaren employees were instructed to greet race victories with the solemn demeanor of a doctor delivering a newborn, Horner commemorated Red Bull's podium finish by plunging into a rooftop swimming pool wearing nothing but a red Superman cape. "There was an energy about us," he says. "We were like a young band on our way up."

The band's first hit came in 2009 with a car known as the RB5. Before the start of that season, the FIA had introduced sweeping new regulations designed to encourage more overtaking during races. The changes covered everything from the bodywork to the tires, while a new innovation called the kinetic energy recovery system, or KERS, was implemented to give cars a short-term power boost on each lap. In sum, the rule changes—which had first been floated in 2006—amounted to a blank sheet of paper for designers up and down the paddock. And no designer was more comfortable staring at a blank piece of paper than Adrian Newey. He picked up his HB pencil, attached a sheet of A4 paper to his drawing board, and began to sketch.

"I do enjoy regulation changes," says Newey. "Perhaps the part of my job I enjoy the most is figuring out what those regulations mean, what is their intention, and if a subtle difference allows [us to explore] new horizons."

The result of those explorations was a Red Bull car unlike anything the team had produced before—with a lower wing at the front, a higher, narrower one at the back, and an innovative "shark fin" engine cover that stretched all the way back to the rear wing. Just as he had in 1998 at McLaren, Newey had digested the rules changes and produced a bold and revolutionary design that went like a rocket ship. Unfortunately, it wasn't quite as bold or revolutionary as someone else's.

That was also the season of Brawn GP's double-diffuser loophole, which proved an insurmountable hurdle. The RB5 was clearly the fastest car on the grid during the second half of the season, but Newey's eye-catching creation couldn't match the downforce generated by Brawn's controversial interpretation of the rules in the first half.

For 2010, Newey worked to upgrade the car, vowing that it would not suffer from a similar deficit next time around. The result was the RB6—which was essentially an RB5 outfitted with a double diffuser. Newey estimates it generated more downforce than any car in Formula 1 history. That meant the RB6 could take some of the highest-speed corners in the sport flat-out. At famous turns like Copse at Silverstone or the long bend on the back straight at Barcelona, the Red Bull drivers didn't even have to lift off the throttle. It was as though the cars were pasted to the track.

Five years after entering the sport, Horner and Red Bull had everything in place. They had the technical expertise, they had the team leadership, they had the fastest car, and, in German wunderkind Sebastian Vettel and Aussie veteran Mark Webber, a pair of drivers each capable of fighting for the championship. Unfortunately, they were more interested in fighting each other.

For all that Horner was the boss of an edgy, upstart team that challenged Formula 1 orthodoxy, there was one area in which he was decidedly old school. Horner steadfastly refused to adopt the No. 1 and No. 2 driver hierarchy that had become standard practice in F1 since Schumacher's success at Ferrari. Perhaps it was his own background as a driver that persuaded him otherwise, but Horner was adamant that once the lights went out, he would let his racers race.

To be sure, there were times when he had to gently remind his drivers that they were in fact coworkers and not mortal adversaries. When Coulthard and Webber had a coming together in the 2007 Chinese Grand Prix, for instance, "he brought us in after the race and his neck was bulging and he was laying into us both about how we've got to play in the interests of the team," Coulthard recalls. "And Mark and I are both taller than him. But at that moment, as he was shouting at us, we knew who was boss."

That approach had served Horner well for his first five and a half years as an F1 team principal. Then, at the 2010 Turkish Grand Prix, the boss lost control. With Webber leading the race and Vettel right behind, Red Bull appeared to be cruising toward a one-two finish. But, on Lap 40, while Webber was in fuel-saving mode, the young German made his move. Vettel edged past his teammate on the long back straight before attempting to cut back to the racing line. The two Red Bulls collided. Vettel was knocked out of the race, Webber sputtered home in third, and a bitter, three-season rivalry was unleashed.

Horner read Webber and Vettel the riot act. This time the bulging neck muscle routine didn't make a difference. Their feud continued for the remainder of the season. When Webber got the better of Vettel at the British Grand Prix weeks later, he shouted, "Not bad for a No. 2 driver!" down the radio.

"The relationship between the two of them was effectively broken from mid-2010," Horner said.

Fortunately, there were no such problems with the car. Red Bull took pole position in fifteen of the nineteen races that year and clinched its first world title at the final Grand Prix of the season. In a single wild race, Vettel swept to victory in Abu Dhabi and topped the standings for the first time all year.

Over the next three seasons, as the relationship between Vettel and Webber grew increasingly strained, Horner grappled with bruised egos and a split garage. But as plenty of his predecessors had learned, having two drivers at each other's throats doesn't really matter when you have the fastest car. Thanks to Mateschitz's investment, Horner's leadership, and Newey's genius, Red Bull had cracked the code.

It consisted of four key elements: aerodynamic efficiency, mechanical efficiency, the engine, and the driver. "If you have good aero and two of the other three, you'll have a good package," chief designer Rob Marshall said. "If you have all four, you'll be unbeatable." Red Bull was sitting on a full house.

The team stomped its way to four successive titles, all won by Vettel, its original protégé. By 2013, Dietrich Mateschitz's team was now so powerful, so successful, and so integral to the future of Formula 1 that it was no longer an outsider. Red Bull was a fully fledged, paid-up member of the establishment. The team that set out to upset the aristocrats now occupied the same rarefied air as those automotive giants.

The clearest illustration of Red Bull's newfound status was the celebration that Horner attended in August 2012 in Switzerland. This one wasn't a race win or a drivers' title. He was best man at Bernie Ecclestone's third wedding.

12

Silver Bullet

MERCEDES HAD ONLY BEEN BACK in Formula 1 for a couple of years when it came to the painful conclusion that this wasn't going well.

At the time they bought the team formerly known as Brawn GP, after the 2009 season, the men from Stuttgart believed they were getting a ready-made championship outfit primed to win right away. The team nicknamed the Silver Arrows was returning to Formula 1 for the first time in more than half a century and the German bosses had made their ambition clear: They weren't coming back for the airline miles. Mercedes had coaxed Michael Schumacher out of retirement after three years and paired him with a young German and son of a former F1 champion named Nico Rosberg. Even for a company whose very essence was aristocratic pedigree and whose slogan was "The Best or Nothing," this was a little on the nose.

So it came as a shock to the executives in Germany when they found that their F1 project was falling into the Nothing category. With two fourth-place finishes in the 2010 and 2011 constructors' championships, Mercedes's overarching trait in the sport wasn't elegance or performance. It was irrelevance. In 1955, the company had quit motor racing altogether following a tragic crash at Le Mans that killed some eighty spectators. Now it was thinking about quitting again because it simply wasn't very good.

The bosses at Mercedes's parent company, Daimler-Benz, needed someone to take a fresh look at their team and explain to them what was going wrong—someone who understood the business and who spoke their language, even if he did it with an Austrian accent. They summoned Toto Wolff.

At that point, Wolff was still years from becoming the debonair prince of the paddock who gave interviews in three languages and traded barbs with Christian Horner. Ahead of the 2012 season, he was a forty-year-old shareholder at Williams, working under Frank at a team that was going nowhere not very fast. Yet he felt he had a handle on the way F1 worked and he was known to the Mercedes motorsport department, because he was an investor in the company's junior driver program and a 49 percent shareholder in the Mercedes-AMG touring car team, which competes in a stock car–style racing series. He could see why Mercedes's Formula 1 effort was spinning its wheels.

Wolff also had a sense that the presentation he had been asked to deliver to Daimler-Benz might be a job interview in disguise, but he didn't get ahead of himself. First, he needed to spell out some home truths. He looked into the numbers and how the team was organized and realized that the main issue was the relationship between Stuttgart and the factory in Brackley.

"Dysfunctional," he says.

Stuttgart thought it should be competing with Red Bull and Ferrari. Brackley looked at the budgets and wondered if it was even competing in the same sport as Red Bull and Ferrari. Mercedes had bought Brawn GP thinking the team was poised for a glorious future. It took Wolff to point out to the company that Brawn GP wasn't built for any future at all.

Even during its title win in 2009, the team had begun running out of money before the end of the season and the performance deteriorated before everyone's eyes. While Red Bull kept upgrading its car, Brawn GP was stuck more or less with what rolled out of the factory in the spring. The double diffuser and the

£100 million golden handshake from Honda had been enough to steal a march on the rest of the pack, but these weren't long-term, sustainable advantages. "All of that was from extraordinary circumstances that made the team win the championship in 2009," Wolff says. "The real status was never communicated."

When he asked the Mercedes execs what their expectations were, he was told, "We bought a world championship team, we want to win world championships." But Wolff had to be blunt.

"I am running on the same budget at Williams," he told them, "and my expectation is to come in the top five. So one of us is wrong."

The Daimler people didn't like that. Dieter Zetsche, the auto industry veteran with a bushy white mustache, barked back that Wolff was talking nonsense. But Wolff threw up his hands: "Don't kill the messenger."

Mercedes took his advice to heart. They hired the messenger instead.

AT FIRST, WOLFF WAS INCLINED TO TURN DOWN THE OFFER from Mercedes. Being a shareholder in Williams allowed him to be the de facto deputy team principal and an owner. So Mercedes had to sweeten the pot. The company told him that it had recently bought back a 40 percent stake in the F1 team from Abu Dhabi's sovereign wealth fund, and if Toto was interested, they would happily sell it to him.

Wolff couldn't resist. "No one had ever been given a chance to co-own a core activity with Mercedes," he says.

Operating the German giant in F1 might have felt strange for the Vienna native, since every other Austrian in the sport seemed to be working for Red Bull. (Even Wolff had worn the Red Bull overalls as a driver in touring cars.) But the relationship with Dietrich Mateschitz, he says, was "love-hate." Mateschitz had been confused years earlier by Wolff's project to list Williams on the stock exchange. How would an IPO help Williams win races? Mateschitz had asked. "I'm not here for the long term. I'm trying to sell shares," Wolff replied. "It's totally different to what you do, Mr. Mateschitz—I'm a venture capitalist."

Besides, Wolff did find one of his countrymen in his new team. Months before he was appointed executive director and managing partner of what was now called the Mercedes-AMG Petronas Formula One team, they had also brought in another investor, Niki Lauda.

What they discovered was that Ross Brawn had played himself into a difficult position. After the genius moves of courting Mercedes away from supplying engines to McLaren, building the success of Brawn GP on them, and cashing out by selling the whole team to the men from Stuttgart, he could hardly make the case that the team needed more money. Brawn had sold Mercedes a winner that appeared to be largely self-funding—and he'd pocketed around $160 million along the way. But the reality was that Brawn GP hadn't spent a dime on developing a competitive car for 2010 and Ross couldn't go asking Mercedes for more backing. Brawn found himself at an impasse. And this time, there was no obvious loophole to wriggle out of it.

"We had 2010, 2011, 2012 in the wilderness," he wrote. "Mercedes had bought the team and were convinced they could run it without investing money into it."

The reason, Brawn believes, is that Mercedes was convinced that the new Resource Restriction Agreement (the cost cap that eventually emerged from the 2009 breakaway attempt) would be enough to keep the team within the budget paid for by sponsors, without a single extra euro from Germany. "However," Brawn adds, "it became clear that Ferrari and Red Bull were not paying any attention to the RRA. And we'd also committed to some fairly expensive drivers."

Wolff had pointed out all of this to the Daimler folks in his presentation before they even offered him the job. And by 2013, it was clear which way the wind was blowing. The outfit that Brawn had built no longer saw him as part of its future—and, if he was being honest, neither did he.

Wolff took over and set about putting his perfectionist stamp on Mercedes everywhere he could. What he lacked in engineering know-how, he made up for in corporate professionalism. Everything the team touched, from the headquarters at Brackley to the powertrains factory at Brixworth to the motor homes it took to Grands Prix, needed to be sleek and pristine. Hand sanitizer went everywhere, long before the pandemic turned everyone else into germaphobes. Garages were cleaned to a standard that even Ron Dennis might have approved of. The bathrooms were reimagined to minimize contact with surfaces and maximize particle airflow.

"In a sport that is engineering-led those values were not prioritized," Wolff says. "Cleaners of a building don't make the car faster. But in fact . . . the feedback I got shows it does. Because it shows your attention to detail. It shows you're seeking perfection. And if you can't keep your building clean, well, how does the car look?

"That's the reset of the standards," he adds.

But the reset couldn't be total. Wolff recognized that two important pieces were already falling into place. Brawn's operational brilliance had identified the sweeping rule changes for the 2014 season, which came with the introduction of new V6 hybrid engines, as a massive opportunity to launch Mercedes to the front of the grid. Working in secret away from the rest of the team for nearly two years, Mercedes engineers had been plugging away on the new-spec engine. Brawn, who'd identified Bridgestone tires and the Honda double diffuser as F1 secret weapons, knew before he left that investing the time to develop the 2014 car, dubbed the Mercedes W05 Hybrid, would produce another game-changer.

And by then, Niki Lauda had helped Mercedes bring in just the person to drive it.

IT MIGHT SEEM STRANGE TO THINK THAT LEWIS HAMILTON'S career was ever in a stall. But late in the 2012 season, the boy-racer prodigy was twenty-seven years old, driving a McLaren car that kept failing him, and beginning to wonder why he was still stuck on just one world championship.

This wasn't where Hamilton was supposed to be. He'd arrived in Formula 1 in 2007 as the surest of sure things, a natural whose instinctive handling, pace, and intelligence behind the wheel destined him for multiple world championships. His team had been certain of this since he was a preteen. The first time Hamilton met Ron Dennis, right after winning the British karting championship at age ten, he told the McLaren boss that he hoped to drive for him one day. Dennis left him his number and said, "Phone me in nine years, we'll sort something out then."

Within three, Dennis was the one picking up the phone. He enrolled Hamilton in McLaren's driver development program, which came with a direct path to a Formula 1 seat if Lewis lived up to his potential. McLaren was going to make all of his racing fantasies come true.

So when Hamilton found himself falling out of love with the team in 2012, it represented a strange and surprising turn in one of the longest relationships of his life. McLaren was the car that Lewis had been desperate to drive ever since he first gripped the wheel of a go-kart as a kid on holiday in Spain. The team represented success, speed—and more than anything else, McLaren stood for Hamilton's hero: Ayrton Senna.

"My dream was always to be like him," he told the *Wall Street Journal*.

Living with his father, who'd been separated from his mother since Lewis was two, Hamilton devoured anything he could about the Brazilian driving genius. He glued himself to a VHS called *Racing Is in My Blood* so often that he wore out the tape. He studied a volume called *Ayrton Senna's Principles of Race Driving* like it was a textbook for its discussion of racing lines and gearing. And on the track, Hamilton learned to mimic Senna's unorthodox, late-braking cornering technique through brutal trial and error. Neither Hamilton nor his dad, Anthony, had ever been around any kind of motorsport before. So as the family traveled to races—a hobby Anthony supported by working three jobs at once and teaching himself to repair go-kart engines—the only thing to do was figure it out on the fly.

Together, they studiously observed the other kids and vowed that Lewis would be quicker. "Through my whole karting career, I never saw any other dad out on the track, doing what my dad did," Hamilton says.

Anthony would stand on the inside of a corner and note where the best kid Lewis's age hit the brakes. Then he'd walk up a few paces beyond that spot and point to the tarmac. "This," he told Lewis, "is where you brake." Hamilton wasn't sure he could do it. Whenever he tried, he'd either spin out or drive clean off the road on the other side. Then, just like that, he developed the touch.

Like Senna, he worked out how to curve his trajectory through a corner as little as possible. Hamilton could carry more speed into the bend, slam violently on the brakes, and get back on the throttle quicker. All of this—every feeling he experienced in the car, every point of technique he picked up—went straight into notebooks that Lewis kept as he built up his personal encyclopedia of racing knowledge. As one of the few kids in go-kart racing from a working-class background, and the only one from a Black family, Hamilton knew that no one was about to give him any free advice. "People saying that, 'You're not good enough, there's no way you could do that, there's no way you're going to achieve anything,'" he remembers. "That adds to that fuel in that fire, 'No, I can.' Karting was my outlet. I realized that when I gripped that steering wheel, that I was connected to this thing, and I could do things with it that those around me seemed to not be able to do. It was like that was my superpower."

After impressing Ron Dennis and receiving around $60,000 to join the McLaren program, Hamilton was on a road closer to what soccer players experience in professional youth academies than most drivers who end up in F1. Other kids, usually from affluent backgrounds, saw their families spend small fortunes on their karting careers until they either flamed out or got good enough to attract sponsorship money. That sponsorship money (often supplemented by a few bucks from daddy) then bought them seats in the lower motor racing series under the pay-driver model. If they showed enough skill there, they might just reach F1. But Anthony Hamilton found it was hard to convince anyone not named Ron Dennis that Lewis was worth investing in.

"He would say, 'How would you like to support the first Black Formula 1 driver?'" Lewis remembers. "My dad would come back saying, 'I've been at all these meetings trying to find sponsorship money but unfortunately, not heard back from anybody.'"

The lack of interest couldn't knock Lewis off the fast track. Four years before he could get a British driver's license, the road to McLaren opened up. Asked by the magazine *Autoweek* in 1998 when he thought he might finally get there, a thirteen-year-old Lewis answered, "Say about 2005."

When he was still in the series just below F1 in 2006, aged twenty-one, Hamilton felt he was about a year behind schedule. But he was getting closer and the McLaren team knew it. For the GP2 sprint race at Silverstone that June, Dennis rented out an entire section of the grandstand along the Hangar Straight. That gave the crowd of McLaren engineers and mechanics a view right down the most hair-raising part of the circuit, a succession of three corners—named Maggotts, Becketts, and Chapel—where Hamilton pulled off a wild maneuver straight off the go-kart track.

Heading into Maggotts, Hamilton closed the gap on a pair of rivals who were already in a wheel-to-wheel battle. By the time they were in the corner, Lewis had attacked around the outside and the cars were running three abreast. That put Hamilton in ideal position to take the next corner ahead of both of them. He had overtaken two cars at once—and showed his future employers exactly what they were getting.

"Yeah," McLaren engineer Mike Elliott remarked to the person next to him. "I think he's gonna be in the car next year."

Once he was actually in that car, Hamilton's first season went exactly as everyone who'd seen him manhandle a steering wheel predicted. Paired at McLaren with Fernando Alonso, the twenty-two-year-old Lewis proceeded to show up his older teammate relentlessly. Not only that, but Hamilton also outperformed

Alonso in the exercise that drivers consider to be the purest test of speed and skill between two teammates equipped with the same car. Over seventeen Grand Prix weekends, Hamilton outqualified Alonso 9–8.

It drove Alonso absolutely crazy. He became paranoid that Ron Dennis was holding back vital data and upgrades to favor Hamilton as his protégé. One former McLaren mechanic even remembers Alonso's entourage handing out brown paper envelopes with 1,500 euros in cash to everyone in the garage, except the crew that worked on Hamilton's car.

Remarkably, Alonso and Hamilton finished the season level on 109 points apiece, a wholly dissatisfying outcome to both men. Even more galling: neither of them was world champion. Amid all of their infighting, they'd somehow managed to fall just one point short of Ferrari's Kimi Räikkönen, who finished on 110.

"I don't think that we were very well managed that year," Alonso says.

No one was surprised when Alonso left the team that winter. Hamilton, untroubled by his own teammate, was free to storm back into a title fight in 2008. The irony is that this time Alonso would unintentionally play a hand in winning him the title. The chaos around Crashgate did just enough to rob Ferrari's Felipe Massa of vital points in Singapore that it opened the door for Hamilton to claim the championship by a point in the dramatic final race of the season.

The way that decisive Sunday in Brazil broke, Lewis needed only to finish fifth or better to become world champion. But with three laps to go and Massa in first, Hamilton was whipping around the wet circuit at Interlagos in sixth behind Sebastian Vettel. There were cars bunched together ahead of them—a mess of backmarkers who'd been lapped plus Toyota's Timo Glock, who was driving on the wrong tires for the wet conditions. Hamilton kept pushing. And somewhere in the confusion and the rain and the freakout inside the McLaren garage, Lewis overtook someone. Not even he was quite sure who he'd passed through the spray of the final corner.

It wasn't Vettel—he'd kept forging ahead and moved into fourth. So as Hamilton crossed the line, it dawned on him that he might have passed Glock. Was that who it was? Was he actually in fifth?

"Do I have it?" Lewis screamed into his radio. "Do I have it?"

Hamilton had it. With a fifth-place finish, he was a world champion for the first time at twenty-three years old and ready to start ripping off title after title. He was blessed with consummate skill, drove an exceptional McLaren, and had now proven that he was the hottest thing F1 had seen since Schumi. It was only a matter of time before Hamilton was on to championships two and three. One day soon, he would equal Senna.

Then came the career stall, like the sound of gears grinding without enough clutch.

SOME OF IT WAS BAD LUCK. THE DEFENSE OF HIS 2008 CHAMPIonship happened to coincide with the Brawn GP season, and the campaigns after that were undone for him by Adrian Newey's latest stroke of genius for a generation of Red Bull cars.

Some of it was McLaren's fault. Ron Dennis was badly diminished by his role in Spygate and the team had let Mercedes slip away as its engine supplier when it partnered with Brawn.

And a lot of it was down to Lewis's inexperience. He was an F1 world champion, but he still tore around circuits with his elbows out, testing the limits of his McLaren and the edge of the rules. Much like his idol, he considered no gap too small to attack—even if the gap was definitely too small to attack. Crashes piled up. The stewards took regular interest. "He drove the car like a go-kart," Mike Elliott said. "It was unbelievable."

In 2011, Hamilton's aggression behind the wheel landed him in fourteen separate investigations by the FIA.

The British press hadn't exactly warmed to him either—a recurring theme on Fleet Street when the subject is young, rich, and Black. The papers had gone after him in 2007, when he moved his primary residence to Switzerland for tax purposes.

And in 2009, the knives were out again after he admitted to lying to stewards over an incident that occurred at the Australian Grand Prix.

On that day, the race was crawling behind the Safety Car following a crash. Hamilton was directly behind a driver named Jarno Trulli when Trulli ran wide around a corner. Hamilton passed him, which is not ordinarily allowed while the Safety Car is out and all drivers are instructed to maintain their positions. Hamilton slowed down to give back the place rather than risk a penalty. But by the time Trulli crossed the finish line in third with Hamilton behind him, McLaren had received word that Hamilton's original overtake was legal. The team promptly appealed, arguing that Trulli had infringed when he re-passed Hamilton. The Italian driver was later docked twenty-five seconds for passing under the Safety Car.

But when the stewards spoke to Hamilton about it, he wasn't entirely truthful about slowing down to give the place back. He was happy to let them think that Trulli had simply attacked and blazed through. "Lewis Hamilton's good-guy reputation takes a battering in 'Lie-gate,'" London's *Evening Standard* wrote. The FIA eventually disqualified him from the race.

For those three maddening seasons, from 2009 to 2011, Hamilton felt he just couldn't catch a break. His finishes in the world championship were fifth, fourth, and fifth, and a second F1 crown now felt about as remote as five-foot-nine Lewis winning an NBA title. "I've had the worst year," he told reporters toward the end of 2011. "If you expect me to be all happy-doolally after a race like that, you're not going to hear it."

In 2012, Hamilton vowed to turn his career around and look for a fresh start. It was time to leave McLaren, the only F1 team he'd ever known. In his biggest move since introducing himself to Ron Dennis as a ten-year-old, Lewis decided to find a car that could win him a world championship, a boss who could truly appreciate his greatness, and a team that could build him a legacy. There was only one place for Hamilton to look.

He went to see Red Bull.

HOW DIFFERENT THE RECENT HISTORY OF F1 MIGHT HAVE been if Hamilton and the team he once wrote off as "not a manufacturer" had been able to strike a deal. "Red Bull are just a drinks company," he'd said in 2011. Then they were nearly his employer.

Instead, Niki Lauda inserted himself into the conversation. In 2012, Hamilton wasn't even sure that Lauda liked him. Lauda, with his three world championships and steely professionalism, had been critical of Lewis's perceived immaturity in the press. But Hamilton agreed to meet him at that season's Singapore Grand Prix. And suddenly, the sixtysomething from Austria and the budding rock star from Stevenage discovered that they had a few things in common.

"You're just like me," Lauda croaked in his accented English.

"Yeah, Niki," Hamilton replied, "I'm a racing driver."

"No, no, no. You're a hard grafter."

Hamilton had come into F1 seeing himself as the boy racer with preternatural instincts, just trying to channel Senna. Paddy Lowe, an engineer at McLaren and later at Mercedes, viewed him as "a clean sheet of paper, a guy that knew this was a fantastic break, and knew that he didn't know anything about racing in Formula 1." Lauda saw him as a grinder.

Once he moved to Mercedes in 2013, Hamilton soon came into his own. He matured—and not just on the track. Toward the end of his time at McLaren, he'd unshackled himself from one father figure in Ron Dennis and another in his actual father, Anthony, whom he'd ditched as his manager. He replaced him in 2011 with XIX Entertainment, the company founded by Simon Fuller, who was responsible for such non–Formula 1 cultural phenomena as the Spice Girls and *American Idol*. The idea was to help Lewis break out of what remained a fairly niche sport and into the right level of global fame.

Running away with the 2014 title at the wheel of a seemingly unstoppable Mercedes certainly helped. That year, the Silver Arrows project that had been cooking since the Brawn days finally got to unleash the hybrid engine it had been developing in secret for more than two years. Under Wolff's leadership, Mercedes claimed victory in sixteen of the nineteen Grands Prix. Hamilton secured the second world championship of his career by winning eleven of them.

When he repeated the trick the following season, it was clear to anyone who watched F1 that Mercedes had put the best driver in the best car, and together, they would be in control of the sport until the technical specs changed again. Hamilton was about to make his case to join the all-time greats.

In the process, Hamilton found that Formula 1 came with a highly specific brand of celebrity. In 2014, Hamilton won the BBC Sports Personality of the Year award—a distinction that no one outside of Britain cares about, often for athletes whom no one outside of Britain has heard about. Yet it goes a long way to capturing the country's sporting obsessions of the moment. Hamilton was preceded by a soccer player, a jockey, two cyclists, and Andy Murray.

Still, being huge in the UK doesn't automatically make you the Beatles. The places where an F1 driver might cause a scene in an airport were located almost entirely in Europe, Asia, and South America. Hamilton understood that real fame—or at least the famous circles he hoped to run in—was reserved for America.

He moved closer to the real A-list in 2015 when he received his first invitation to the Met Gala, the fashion world's annual parade of who's hot and who's hip, organized by *Vogue* at the Metropolitan Museum of Art in New York. Hamilton made his way there via Las Vegas, where he watched Floyd Mayweather Jr. fight Manny Pacquiao. And in a campaign to promote a British brand hoping to make the leap into couture, he showed up flanked by a young Emily Ratajkowski and Bella Hadid.

By the following year, Lewis's cachet was clearly on the rise. He was back at the Met Gala, this time wearing Dolce & Gabbana, and rubbing elbows with Kanye West, Kim Kardashian, and Beyoncé. Except there was a paradox about Hamilton's skyrocketing celebrity. He was young, charming, and riding high at the peak of a glitzy sport, which made him hugely popular among other famous people. He developed close, very public friendships with the likes of Rihanna, Justin Bieber, and Serena Williams that played out on Instagram and behind velvet ropes. Hamilton found that he moved easily on a uniquely American scene that celebrated Black excellence—a nexus of music, fashion, and sports that simply didn't seem to exist back in Europe. He was growing into the A-lister's A-lister. Everyone inside the building at the Met wanted to have their picture taken with him. Yet if Lewis stepped outside onto Fifth Avenue, he could still walk around New York unbothered. To America at large, F1 famous just wasn't that famous.

Hamilton didn't feel like he completely fit in in Europe either. His arrival as the first Black driver to compete in Formula 1 had been met with vicious racism that he never forgot. (An African American racer named Willy T. Ribbs had been the first Black man to drive an F1 car when he tested for Bernie Ecclestone's Brabham in 1986, but he never took part in a Grand Prix.) In Spain in 2008, one group of fans showed up in blackface and T-shirts that read "Hamilton's Family."

"No one said anything," he says. "I saw people continuing in my industry and staying quiet."

Hamilton was constantly reminded that he didn't fit the historic mold of an F1 driver. And even as his status rose, he never shed the chip on his shoulder. The same kids he'd seen in go-karts with dads who didn't have to work three jobs to pay for it were now all around him in Formula 1. One of them, Nico Rosberg, was even his teammate. The only person driving the same world-beating equipment as Hamilton happened to be the German son of a former F1 champion who'd grown up in Monaco. By 2016, any veneer of friendship between the two had peeled away.

That made the season extra awkward for anyone staffing the Mercedes garage. The Silver Arrows won nineteen of the twenty-one Grands Prix that year, with Hamilton taking ten to Rosberg's nine. But two things gave Rosberg the chance to leapfrog his teammate. One was doing a better job of damage limitation in the races that he didn't win, with five second-place finishes and two thirds. The other was taking advantage of four major engine failures on Hamilton's car throughout the campaign. The tension became exhausting for everyone.

No one "has been able to manage two No. 1 drivers battling for a championship more than a year or two," Toto Wolff told the *Wall Street Journal* as the season reached the business end. "We are in our fourth year."

He knew that no matter which way things shook out that season, there wouldn't be a fifth. "Hopefully in 10 or 20 years when we look back, people will say 'They've kept it together quite well,'" he said.

The truth was that they hadn't. Wolff could sense the obvious chill between the two sets of mechanics and engineers who worked on each man's car. Relations only grew frostier as Hamilton closed the gap on Rosberg all the way to the last Grand Prix in Abu Dhabi. Yet despite finishing second to Lewis in each of the final four races, Rosberg held on to match his father, Keke, and win a world title. Knowing how close the battle had been, even when Hamilton had been plagued with bad luck, Rosberg decided he didn't feel like trying again. Within forty-eight hours, Nico announced his retirement. He was just thirty-one.

"I'm just really glad it's over," he said.

With Mercedes being so dominant, a little intrasquad tension was about all the drama Formula 1 could muster in those days. Even Wolff had to admit that it was a risk for the sport. The most he could do was let his drivers race each other and not box them in with team orders. But the reality was that unless you really liked Mercedes, there wasn't much point in watching anymore.

The second Rosberg left, Wolff ended the experiment of not declaring a clear No. 1 driver. He soon brought in someone a little more comfortable with playing second fiddle, a Finnish company man named Valtteri Bottas. That freed Hamilton up to go right back to mopping the floor with the competition. From 2017 through 2019, he won exactly half of the sixty-two Grands Prix. For anyone who remembered the Schumacher years, things were beginning to feel eerily familiar—so much so that Fernando

Alonso would announce in 2018 that he planned to quit the series for a while and explore Indy or Le Mans, because Formula 1 was too predictable.

"I can't sell tickets for a shit product," the promoter for Silverstone complained in *The Independent* at the height of the Mercedes dominance. "I've said that people don't come to watch guys looking at data screens . . . How long is it before the technical director is stood on the top step, not the driver?"

Silverstone's ticket sales might have suffered, but even with a completely unassailable Mercedes, the overall business was in better shape in 2016 than it had been in ages. Despite a dip in ratings to just 1.52 billion viewers for the entire 2015 season, the F1 Group posted a profit in 2016 and CVC Capital Partners was able to pay the teams five times what they were doling out in 2006, thanks to an agreement to share all F1 revenues with them instead of just the income from broadcast rights. It was enough to make the company start to feel that it had held on to this asset long enough.

A decade around Formula 1 had taught CVC that the sport rarely stayed this quiet. Even if things were uneventful on the track, a period of financial stability and minimal scandal might just be the moment to call it a day.

Luckily for them, there was no shortage of interested parties who saw F1 as much more than a series of often boring events. Rather, they viewed it as a sleeping giant with built-in glitz, a proud history, and the kind of global reach that most other sports would kill for.

Bernie Ecclestone had built it up since the 1970s as a purveyor of TV rights and sponsorship, an intermediary with circuits, and a mediator between the teams and the FIA, all for a hefty fee. But for whoever took over the series from CVC—and from its most ancient employee—the essence of the product would boil down to something far simpler. Formula 1 in 2017 needed to be treated as the entertainment product it really was: a prestige television drama.

13

Changing of the Guard

IT SEEMS ALMOST TOO PERFECT that the sequence of events that irrevocably changed the modern history of Formula 1 should begin on a chartered yacht bobbing in the harbor of Monaco. But after months of courting buyers for their motor racing series, CVC knew that a little cliché could go a long way.

So in the spring of 2016, after wading through dozens of potential bidders—some serious, some not so serious—CVC's Donald Mackenzie invited a small group of executives from a company called Liberty Media, along with their wives, to join him and Bernie Ecclestone for a classic Grand Prix weekend and see the very best of the prized asset he was putting up for sale. After a decade of owning F1, Mackenzie felt that it was time to cash out.

Private equity companies don't tend to get rich by sticking around forever, and ten years was twice as long as CVC typically held on to anything. F1 had been a rare exception. Despite a failed attempt to list F1 on the stock exchange in 2012, CVC had successfully sold off various pieces of the company to earn some $4.5 billion from its initial stake of $1 billion. Now it hoped that the 35 percent stake it had left would be worth at least that much again.

Finding an interested party hadn't been a problem. Dozens of suitors picked up the phone imagining that they could be the ones to maximize the potential of an already global sports league that was crying out for a refresh. CVC hadn't been a popular owner with fans—the perception was that they had taken money out of the sport rather than reinvesting it—but what did they expect? This was private equity, not the F1 charitable foundation.

Besides, CVC had the distinct impression that it had carried the sport about as far as it could. There were still more markets to conquer, namely the United States, where Mackenzie had a vision to add races and claw back market share from NASCAR. And some serious work was required to update the way F1 used new media. But CVC realized that those projects would have required years more effort, and by now, Mackenzie was ready to sell.

The new task became sorting the serious suitors from the pretenders. Stephen Ross, who owned the Miami Dolphins and had made a fortune bringing European soccer teams to the United States, took a long hard look, as did Ari Emanuel, the CEO of the William Morris Endeavour entertainment powerhouse. Not for the first time, Rupert Murdoch was rumored to have considered a bid. But Mackenzie quickly took a liking to the executives from a company called Liberty Media. They checked all the boxes. Founded by the telecom magnate John C. Malone in 1991—a man who had once held nearly a third of News Corporation stock and worried Murdoch that he was coming for the whole company—Liberty had grown out of the rapid deregulation of US cable television and spread its influence into every facet of American entertainment and communications. At various points, it held stakes in radio stations, local TV stations, Barnes & Noble bookstores, and the Sprint cell phone networks as Malone imagined the endless entertainment offering of what he called the "500-channel universe." And by the end of 2016, Liberty's market cap was approaching $12 billion.

The company's only involvement in major sports at that point was owning the Atlanta Braves baseball team, which had about as much in common with Formula 1 as its chain of bookstores. But that's not to say they were clueless. What put Liberty over the top in CVC's eyes was that the Americans came with an executive whose reputation preceded him as distinctively as his handlebar mustache. He was a cable television veteran and Murdoch protégé named Chase Carey, who had been chairman of 21st Century Fox and, just as important, helped launch Fox Sports. Carey would be flanked by industry veterans Greg Maffei and Sean Bratches. Hoping that these might be the guys, Mackenzie spent the Monaco weekend schmoozing them and introducing them to Justin Bieber. He paraded them around the paddock to show them the loudest, sexiest side of what he was pitching. The principality had been used to sell the sport to generations of fans; now it was being used to sell the business to a couple of Americans.

If everything went according to plan and they could agree on a price, these were the men who would take Formula 1 off CVC's hands. Once they did, Mackenzie viewed Carey as someone who could eventually step into Bernie Ecclestone's shoes. Never mind that Carey didn't see himself that way—at least not back then.

THERE WAS HARDLY A MOMENT DURING CVC'S YEARS OF OWN-ing Formula 1 when Bernie didn't have some sort of legal fight hanging over him. Most of the time he was able to brush them off or spend enough money to make them go away. But by the 2010s, the whiff of scandal was getting harder to shake, and it was beginning to become a problem for his private equity bosses.

That's when Ecclestone suddenly found himself on the wrong end of the most expensive case of his life. It centered not on a team principal, or on a manufacturer, or even a scorned promoter. His surprise nemesis this time was a German banker named Gerhard Gribkowsky, and he was making some rather alarming allegations.

Gribkowsky had been the chief risk officer for Bayerische Landesbank, which, you might recall, inherited a large chunk of the Formula 1 business in the early 2000s and later sold that piece to CVC. But in 2013, a Munich court found that Gribkowsky's recommendation to sell to CVC had in fact been made around the same time as a payment of $44 million from Ecclestone, which came partly in the form of a personal check signed by Bernie himself. So Gribkowsky was sentenced to eight and a half years in prison for fraud and embezzlement. Ecclestone, the court ruled, had unduly influenced Gribkowsky to favor a sale to his preferred buyer because CVC had committed to keeping Bernie around as chief executive.

To the outside world, the whole affair looked an awful lot like a bribe, which Ecclestone denied. But that didn't stop other former F1 shareholders, including a German company called Constantin Medien, from getting involved and calling it what it was. Suing in British court, Constantin alleged that the "corrupt agreement" between Ecclestone and Gribkowsky led to Formula 1 being undervalued in the sale, costing them more than $150 million.

Ecclestone never said he didn't make the $44 million payment—the evidence of it was out in the open. But he suggested that the real victim in all of this was actually him.

When prosecutors in Munich saw the London case and took an interest, they dragged Ecclestone before a judge in 2014, where Bernie testified that Gribkowsky had actually been shaking him down with a threat to go to the British taxman over Ecclestone's secret control of a family trust. Ecclestone added that he would have been on the hook for $2 billion in taxes. After several months of litigation, in both London and Munich, the final conclusion was that Ecclestone had managed to dance between the raindrops again—although a little more expensively than usual.

The London court ruled against Constantin Medien, finding that Ecclestone's actions hadn't caused the company to suffer a financial loss, while German prosecutors failed to secure a conviction and the court allowed Bernie to walk away with an official status of "neither convicted nor acquitted" in exchange for $100 million—a common process in the German legal system. Ecclestone paid the fine through gritted teeth. At least it meant that he could get back to work. During the drawn-out proceedings, he'd been obligated to hand the reins of F1 to his longtime legal counsel, a Cambridge-educated lawyer named Sacha Woodward-Hill, whom he'd personally recruited in 1996. Now the eighty-four-year-old Ecclestone could concentrate on motor racing again. Only by Christmas, he clearly wasn't over the whole drama.

Bernie's 2014 holiday card featured a cartoon of himself dragging a sackful of money through a snowy German street with the words "$100 million" on it. A masked bandit was holding him at gunpoint.

"This is not a robbery," the gunman said. "I am collecting for the Bavarian state."

The entire episode was costly, stressful, and deeply embarrassing for CVC, but nothing they weren't used to in a decade of working with Ecclestone. The other, longer-term problem they had with him was more strategic as CVC began to consider life after F1: Bernie was hopelessly out of touch. The model he'd spent four decades building for the sport no longer made as much sense as it once did, especially not when Ecclestone was so ill-suited to handle the two areas where CVC felt F1 still had room to grow.

The first was the American market—they knew how Bernie felt about that. Parking lots in Las Vegas, ostrich festivals in Phoenix, and exploding tires in Indianapolis had done plenty to turn him off the United States. The only new territories he wanted to explore all lay east of the Princes Gate offices.

The second issue was F1's need to embrace the new media landscape created by the internet, social media, and everything else that people younger than eighty-four treated as essential. In this department, the sport was way behind. It tweeted, but not well. It had a website, but it wouldn't sell a banner ad for it until 2018. As far as Bernie was concerned, reaching a younger audience wasn't a priority because younger audiences, he said, "don't buy Rolexes."

One young Formula 1 lover took particular exception to this approach. All he wanted to do was share F1 content with his large social media following without running afoul of Ecclestone. But every time he posted a photo or a video, he seemed to get another cease-and-desist letter from the lawyers at Princes Gate accusing him of illegally distributing their intellectual property. That F1 fan's name, by the way, was Lewis Hamilton. And even he could tell that any sport pretending that its biggest stars weren't the central actors of the whole show had no chance of reaching new audiences in the twenty-first century.

Under Liberty, it would be a different story. Months before the sale went through, Sean Bratches was in London making plans for the company to revamp the image of the sport entirely. A former lacrosse player at the Rochester Institute of Technology, he'd joined ESPN in its freewheeling early days of the 1980s and wound up staying for three decades, until he was hired away by Carey and Maffei in mid-2016. In Bratches's view, F1 "wasn't punching at its weight class." He could see the reach and history of this series, yet couldn't understand why no one around him ever seemed to be talking about it.

On one of his first days in the UK, Bratches walked into the local offices of Wieden+Kennedy, the legendary ad agency that had worked with all kinds of sports figures looking to sell the world a new narrative, from the Brazilian national soccer team to Lance Armstrong. He commissioned a global brand study for Formula 1. Bratches needed to know exactly who was watching this thing and, more important, who wasn't. With two dozen focus groups spread across four continents, bringing together casual fans, avid supporters, and people who barely knew what F1 was, Wieden+Kennedy conducted hours of interviews to produce a dense report into the state of Formula 1 viewership in 2017.

One thing stood out above all else. Fans viewed F1, Bratches said, as "impenetrably exclusive." And who could blame them?

The experience of watching a race often came down to listening to middle-aged men debating how quickly some rubber was falling apart. Anyone expecting displays of pure speed often tuned in to hear discussions over arcane rules instead. And for all the talk by F1's sponsors of creating an "aspirational" image for the sport, it turned out that there were only so many Rolex ads and parades of ostentatious wealth that viewers were prepared to take before deciding that this wasn't for them at all. Simply put,

Formula 1 was sitting on something potentially great, but it had also become its own worst enemy.

Bratches's takeaway was that for too long, F1 was being run by Bernie and the teams for the exclusive benefit of Bernie and the teams. This series needed to focus on the consumer again, and Liberty's job was to "unleash the greatest racing spectacle on the planet."

If that felt a little bit too much like Madison Avenue corporate sloganeering, then buckle up. This was the man who'd developed the catchphrase "This is *SportsCenter*" during his ESPN days. In hoping to understand what was wrong with F1 and then fix it, Bratches soon identified what he called five "key behaviors" for the sport and its fans.

They were, in no particular order: "Revel in the racing," "Make the spectacle more spectacular," "Break down borders," "Taste the oil," and "Feel the blood boil." Let those sink in for a moment.

Without getting too caught up on the taste of motor oil, those goofy slogans hit on a couple of larger truths. Liberty was really saying that F1's central problems were that it was boring and inaccessible to the audiences it was looking to develop, both geographically and in the way it was distributed. Bernie's preference had always been to run the sport as a closed shop so that he could control (read: monetize) every ounce of F1 content that dribbled out into the world. Bratches says that when he first had lunch with the sport's biggest star in March 2017, Lewis Hamilton actually brought the stack of cease-and-desist letters he'd received over the years for sharing behind-the-scenes images from Grands Prix on his Instagram.

That would have to change. So would plenty of the sport's more obscure regulations, provided Liberty could learn how to manage the complex relationship with the FIA. Too much of the action was being decided off the track and turning off more casual viewers.

"We have got too many complicated penalties and rules," Chase Carey would later tell an investor conference. "We have got a 100-page regulation book. We have got to get the business to a place where it is easier to follow and has fewer complexities that fans out there really can't follow.

"It will always be a complicated sport that is a marriage of sporting competition and technology," he went on, "but we need to make it something that is more in line with what the fans want to see and what excites and energizes them."

Nothing was too entrenched for Liberty to dislodge—or at least try to. To them, Formula 1 was just another set of broadcast and audience problems to be tackled. And despite being an enormous media group, they talked the talk of wanting to run F1 like a start-up. What had for decades been the world's immovable, unquestioned premier motor racing series was being reframed by the Americans as something of an underdog story. No longer would it be a sport run by billionaires designed entirely to hawk fancy products to a general audience that might never even recognize a Rolex.

Somehow, Formula 1 had grown from a working-class sport built by petrolheads for petrolheads into a global luxury catalog and gotten stuck in that state. Even when its popularity declined in the twenty-first century, F1 kept acting like the snooty sales assistant on Rodeo Drive, pretending that anyone who wasn't already wearing a fancy watch and driving a Ferrari could never become a meaningful customer. If you had to ask how much it was, you couldn't afford it.

Liberty understood that F1 needed to get over itself.

It would still be run by billionaires, obviously, but now they'd be meeting audiences halfway, in places like esports, social media, and, though they didn't know it yet, Netflix. The entire project was about making their world a little more likable. But in order to do that, they needed to adjust a few perspectives inside the sport.

After nearly seven decades of teams fighting tooth and nail for every advantage, every dollar, and quite literal survival from season to season, Liberty presented them with a new reality. Instead of being rivals, these teams had to understand once and for all that they were all in business with each other. It wasn't enough to come together once every so often to extract a larger share of television or commercial payments out of Ecclestone during Concorde Agreement negotiations. The popularity of F1 depended on their realizing that grinding each other into dust didn't help anyone. The resource restriction agreement put in place under CVC had begun the work of making F1 less financially ridiculous, but the sport needed more belt tightening—even if this fundamentally altered the essence of an enterprise that had never been about acting reasonably. What had once been a free-for-all of teams spending to put themselves at the absolute vanguard of technology would have to become about doing their best within the framework of an ever-shrinking cost cap.

The sport that used to be about going as fast as humanly possible would soon be about going as fast as was financially prudent. Besides, Liberty knew it wasn't in the business of selling pure speed—"Revel in the racing" was only one of the five key behaviors. As any American in the entertainment business knew, it was also about selling personalities.

THE IMAGE OF FORMULA 1 THAT LIBERTY WANTED TO CREATE was gradually coming into focus. Over the months of protracted talks, they had learned everything they could about the business, despite being under strict instructions to go through the diligence process in the tiny meeting rooms of Princes Gate without removing or photocopying a single document. But at least they had a plan. And against all expectations, Ecclestone was more or less playing nice with Carey.

"We have somebody that knows what they're doing for a change," Bernie told a clutch of reporters one day as the pair walked out of Princes Gate together.

"I'm just learning from Bernie," Carey replied.

Carey would soon demonstrate just how much he'd picked up from the old ringmaster. In September 2016, Liberty announced that it agreed to a deal to acquire the series for $4.4 billion and take over all of its debt. Carey would become chairman with Ecclestone remaining as CEO. The plan was to hand Bernie a three-year contract—until it suddenly wasn't. Ecclestone, who had spent forty years keeping people on their toes by never quite letting them in on his thinking, found himself in the dark while Malone, Carey, and Maffei wrapped their heads around the business and adjusted the terms of the deal. And sometime that winter, the new owners at Liberty changed their minds about Bernie. The more they looked at F1, the less they felt Ecclestone had to offer as they refashioned this middle-aged sport to capture younger audiences, America, and the internet. The octogenarian key man of Formula 1 was no longer mission critical.

On Sunday, January 22, 2017, the day before Liberty prepared to announce that its full takeover was complete—including the acquisition of the minority stake that they originally planned to let Bernie keep—Carey rang Ecclestone to schedule a meeting: 10 a.m. Monday morning, Princes Gate.

By the time Bernie walked through the doors, he had a sense of what was coming. He sat down across from Carey and the two men skipped the pleasantries.

Liberty had acquired the company, Carey began, and there were going to be some changes, Ecclestone recalls. Top of the list was appointing a new CEO.

"I want your job," Carey told him.

Not for the first time in his life, Ecclestone had been outfoxed. The difference now was that there was no way back. His old lieutenant, Woodward-Hill, appeared in the room with a stack of papers: it was his official resignation from Formula 1. More than forty years after he'd first dabbled in a ragtag circus that fancied itself the peak of motor racing, Bernie was handing over the keys to the empire he'd built. His shares were worth several hundred million dollars, if that was any consolation. But for once, Ecclestone wasn't thinking about money.

"They were determined that anything to do with me was going to have to go," Ecclestone says. "They thought, basically, 'Here's an eighty-year-old guy that's been running his bloody sport for forty-odd years and I'm sure we can do a lot better.'"

Stony-faced, he asked for a pen, grabbed the papers, and signed them, as usual, without reading a word. In prewritten statements, Liberty thanked him for his contributions and Ecclestone discovered that his title was now chairman emeritus (although the proper Latin term was really persona non grata). "I'm proud of the business that I built over the last 40 years and all that I have achieved with Formula 1," read the press release quote Liberty had drafted for Ecclestone, before going on to thank the teams, sponsors, and broadcasters. But in private, Bernie was far more to the point.

"You've bought the car," he told Carey after signing over control. "You might as well drive it."

14

Magic Sauce

"WE ONLY KNEW BERNIE-WORLD," SAYS Toto Wolff.

No one had ever confused Bernie-World, essentially a high-octane amusement park that stretched out over five continents, for the Happiest Place on Earth. But at least they knew their way around this version of F1. They knew Bernie's track record, they knew his foibles, and they knew how to argue with him. Formula 1 was far from a perfect business, but it just about held together. The teams were making more money than they ever had, thanks to CVC. The sport had opened up new territories, from Istanbul to Mexico City. And for all of the petty disputes, the whole thing was functional. It wasn't the most forward-thinking organization, yet after more than four decades, the triumph of Bernie Ecclestone had been to keep everyone in the same room and make all of them rich enough to keep racing. Sure, there were times when nearly every person in Formula 1 wished that he would just retire already. Now it had actually happened. The five-foot-two devil they knew suddenly looked a lot less scary than the corporate devil they'd never met.

"How can you question the individual who's built all that, who made us what we are today?" Wolff adds. "Then you have a new owner: Americans who have no idea about the sport, but a lot of ideas about media rights."

Despite his deep skepticism—one of the few things on which he could agree with the other team principals on the paddock—Wolff couldn't fault the Liberty executives for their enthusiasm. Early in the post-Bernie era, he met with Sean Bratches, the company's new commercial czar, and listened to his pitches for bringing F1 to a larger, younger, more online audience. Bratches spoke about harnessing the wealth of new platforms available to take fans deep into the paddock. He envisaged a vehicle that would capture the stars of the sport up close and unfiltered, treating them like the international celebrities they really were. Wolff thought he knew where this was going.

Bratches's revolutionary plan, which no sport had ever tried successfully, was for F1 to operate a golf-cart train that ran through the paddock and past the team garages. The inner sanctum that Ecclestone sought to protect at all costs—except from those willing to pay the extortionate rates of the Paddock Club and other corporate hospitality—was about to become something like the tours at NASA or on Hollywood studio lots. Wolff saved his thoughts on this idea until he left the room.

On his way out, he turned to the Mercedes colleague alongside him.

"What was *that*?"

BRATCHES WOULD COME UP WITH OTHER, NON-GOLF CART train ideas, and in truth, the reactions from inside F1 weren't much more enthusiastic. Even the plan that would turbocharge the sport and make Liberty look like geniuses was met with profound reservations. But after speaking to an old associate of his from ESPN who was now working for Netflix, Bratches was certain that this scheme could work.

F1 was going to commission a ten-part, fly-on-the-wall reality series about itself. In a sport defined by perpetual reinvention, no one had ever dared to touch the sanctity of the pit lane. This was where the real business of racing unfolded and the one place where the teams were truly in charge. What happened in the garages stayed there—they were the last sacred domain in F1. Now Bratches wanted to jam film crews in there.

He couldn't force the teams to participate in this project, provisionally dubbed *Drive to Survive*, but he hoped that they would see the benefit of doing the one thing that Ecclestone had spent all that time trying to prevent. Eight of them signed on right away, recognizing that in their mission to give sponsors as much visibility as possible, more screen time was worth a small sacrifice in secrecy.

The two outfits that held out, Ferrari and Mercedes, were the ones that didn't need any help getting screen time. In fact, Mercedes was already working on its own deal for a documentary with Amazon. And Ferrari refused because that's what Ferrari always did. As Bratches puts it, the culture in Maranello was very much, "No. Now what was the question?"

Privately, both teams were also appalled at how little Netflix was paying for access. While broadcasters around the globe were shelling out tens of millions of dollars to carry the races live, the world's most popular streaming service was gaining an unprecedented peek inside the sport for "a few million," according to one person briefed on the original deal.

The other eight teams on the paddock had no such qualms. The crew from Box to Box Films, the production company behind the series, landed at the Brazilian Grand Prix in late 2017 to have a nose around. And the two producers, James Gay-Rees and Paul Martin, still had no idea how *Drive to Survive* might look. They were supposed to begin shooting in three months.

The tricky part about pulling back the curtain on Formula 1 is that the very things that make F1 appealing to traditional fans are inherently difficult to capture on film. The chess match around pit stops and race strategy is incomprehensible to casual viewers. The course of entire seasons is determined in highly technical meetings that take place in factories in the middle of the British winter. Even the drivers spend the whole time with their Hollywood looks obscured by clunky helmets that make them appear like identical bobblehead dolls.

Because cameras had never been where the *Drive to Survive* crews were going, much of the paddock wasn't sure how to behave. Plenty of mechanics couldn't stop themselves from staring straight down the lenses. "It was distracting," says Lewis Hamilton, who had been followed by film crews for his entire professional life. "All my engineers, they're like, 'Oh, there's a camera on me.'"

But once the previously anonymous worker bees in the paddock got used to the cameras in their faces and boom mics hanging over their heads, a few surprises emerged. What even the producers hadn't realized was that, behind the veneer of corporate sponsorship and robotic good boys who are born to sell watches, F1 drivers and team principals were actually world-champion shit-talkers and prima donnas. At this point, they understood that they weren't making a straight documentary or reportage. It couldn't be Formula 1 meets *60 Minutes*.

This was *The Real Housewives of Monte Carlo*.

"Sport is the original reality television," Bratches says.

The individuals who grasped this instinctively became the stars of the show right away—especially without Ferrari and Mercedes there to take up all the oxygen in Season 1. Guenther Steiner, the Italian engineer who had no business being famous and was once cut loose by Red Bull, delighted audiences with his foulmouthed reactions to the misfortunes of his sad-sack Haas team. ("Now we are a fucking bunch of wankers," is one of his many contributions to the F1 lexicon.) Christian Horner, previously just another Brit on the pit wall, revealed himself to be one of the most viciously cutting characters on television since Simon Cowell, embracing his role as a heel by needling his rivals to amuse the viewers. And Daniel Ricciardo, the permanently shirtless Australian Red Bull driver, almost never won races, but drew in hordes of new fans with his million-dollar smile and goofball nature.

That star quality stood in stark contrast to the actual star of the team, a brooding Dutchman named Max Verstappen. This was the kid who was practically raised in the paddock, the son of journeyman F1 driver Jos Verstappen and a former go-kart racer named Sophie Kumpen. By the age of sixteen, he was already in the Red Bull system and earned the role of test driver for the Toro Rosso team. And by seventeen, he was on the F1 grid, the youngest driver ever to start a Grand Prix. (The FIA later ensured that his record would stand for years by raising the minimum age for the super license required to drive in Formula 1 to eighteen.)

Not since Lewis Hamilton had the sport seen such an obvious natural. Max's teenage impertinence and aggressive tactics called to mind a young Lewis. If Hamilton was the obvious successor to Ayrton Senna, the brash Verstappen, who had a habit of shouting back at his engineers and an aversion to instructions, was closer to Michael Schumacher. He had no interest in fashion, red carpets, or anything vaguely resembling global celebrity, unlike the man who'd become his fiercest rival. If he was being honest, Max would rather be at home in front of his computer playing F1 video games. Which is perhaps why the camera loved the outgoing Ricciardo a little more than the Red Bull driver who actually won races.

Because *Drive to Survive* wasn't an old-fashioned documentary, it didn't have to play by the rules of one. Conversations weren't scripted, but they weren't always natural either. A question or two might be planted. The editing might occasionally veer toward the sensational. Breaking the fourth wall became a recurring feature. It didn't matter. This was reality TV gold.

"I hated it at the beginning," Wolff says. "On a flight to Australia, I watched a few episodes of Series 1—a nightmare. Hollywood in Formula 1."

There had been plenty of other series that purported to go inside the locker room. *Hard Knocks* had exposed the inner workings of NFL teams since 2001, and Amazon had turned that kind of lightly varnished sporting advertorial into its *All or Nothing* franchise. The difference with *Drive to Survive* was that the series wasn't just embedded with one team. It could mine the entire sport for storylines and rivalries. The producers caught such unexpected footage that they had more than they knew how to use. No one had expected the highly secretive world of F1—one that had almost collapsed under the weight of a literal espionage scandal—to be so candid.

"Not many shows are made where you've got hugely successful European businessmen and billionaires getting upset about front wings in a car park in Northamptonshire," says *Drive to Survive* producer Paul Martin.

Not many sports shows are able to drum up interest in the teams that never win, either. But Netflix made heroes out of the strugglers at the back of the grid and street fights out of battles for ninth place. At the peak of F1's popularity in the 1980s and 1990s, even avid fans struggled to name the seemingly interchangeable drivers of the second-tier outfits. Now they knew their life stories as they bared their hopes, dreams, and radio freakouts in ten episodes a year, crafting drama out of race results that were several months old. *Drive to Survive* explicitly rejected the notion that finding out who wins was the central purpose of watching sports programming. Repacking stale competition worked because the actual outcome was irrelevant.

Even more surprising than the teams' openness: the show turned into a hit. Though Netflix guards its numbers more fiercely than a new sidepod design, the participants could tell in their daily lives that they were reaching a new audience. Their friends, relatives, and neighbors not only had sudden thoughts about teams in the middle of the pack and driver attitudes, they also wanted tickets to Grands Prix.

"It's fast-tracking people through the years," Hamilton says. "I'm getting people reaching out to me, who also had never watched and are crazy fans now."

The show was extremely watchable on its own, with its mix of hand-holding for new fans, dramatically lit confessional interviews, and a wealth of in-garage and on-track footage from dreamy locales. But what no producer in the world could anticipate was just how perfectly *Drive to Survive* would nail its timing.

Season 2 of the show, now featuring Mercedes and Ferrari, dropped on February 28, 2020. You may remember what happened next. A whole bunch of people who had never shown the slightest interest in Formula 1 were about to spend a lot of time at home, scrolling through Netflix.

Formula 1 wasn't immune to the pandemic—and the realization dawned on the sport quicker than most. Teams and officials had flown halfway around the world to Melbourne for the first Grand Prix of the season in mid-March, just as the headlines about something called the "novel coronavirus" were becoming panic-inducing. But until a few hours before the first Friday practice was due to begin, F1 was prepared to race. And then it suddenly wasn't. Without running a single lap in anger, the teams packed up their gear and left Australia, unsure when they might get on track again.

Four months passed without a Grand Prix. But through *Drive to Survive,* Formula 1 could sense that it was turning into one of the obsessions of the early pandemic, right up there with sourdough bread and attending work meetings with no pants on. Netflix's breakout hits of 2020 somehow included shows about private tiger zoos, a fictional chess prodigy, and a seventy-year-old motor racing series. Against all odds, F1 had carved out its own corner of the weirdest media landscape anyone could remember. The sport simply couldn't waste this chance.

So F1 undertook a massive effort to cram in a seventeen-race schedule into dates between July and December. Monaco, Vietnam, Canada, and the United States all dropped off the calendar, as the sport designed back-to-back weekends with two races held in the same location, inside virtual bubbles. (This wasn't much of a stretch for F1, since each Grand Prix was effectively its own bubble to begin with.) They ran on consecutive Sundays at the Red Bull Ring in Austria, at Silverstone in the UK, and at Sakhir in Bahrain. The rest of the time they traveled wherever local Covid rules allowed. But F1's efforts to stay on television in front of a global captive audience paid off. By the end of 2021, the sport had added some seventy-three million new fans in its ten biggest markets, according to research by Nielsen. Clearly, something was working and Netflix was at the heart of it. Even F1 insiders had to admit they were surprised. Anyone who said they were certain that *Drive to Survive* would turn into a global hit was simply lying.

"We have no idea if it was a genius move or a lucky punch," Wolff says. "We can't really understand which of the ingredients gave the magic sauce."

THE MOST REMARKABLE THING ABOUT THE SPORT'S SURGE IN popularity through a highly bingeable TV show is that it expanded the definition of what it meant to be an F1 fan. There were devotees of the show who now needed to vacuum up as much content as they could and happily committed to hours of weekend race viewing—even if a two-hour Grand Prix still had a nasty habit of being technical and boring. Then there were those fans who couldn't care less about tire strategy, or even who finished where. They wanted Toto and Christian spilling tea, Lewis modeling Parisian couture, shots of celebrities on yachts, and Danny Ric kidding around on Insta. The hierarchy of the sport itself barely changed. Mercedes was still way out in front—so much so that its dominance became merely a subplot—the smaller teams were still hopelessly scrapping for a handful of points, and Ferrari was still finding new ways to exasperate its fans.

"We have different audiences we need to respect," says Stefano Domenicali, the former Ferrari boss who was installed as F1 CEO by Liberty in early 2021. "There is the audience that is interested in our lifestyle. Now, F1 is perceived as the place to be . . . So we need to talk with this language. Then we have the fans that are into super detailed analysis on the technical side of it. We need to give them services, content, and analytics. And then there is another new community that is talking about Formula 1 from a business perspective . . . That's the beauty of where we are today."

Inside the Liberty F1 offices, they imagined all of their new fans in a funnel—classic consultant-speak—with things like *Drive to Survive* and drivers' social media at the lip, and the real business of racing at the bottom. The idea was to find ways to move people through it. But it wasn't long before F1's head of broadcast

Ian Holmes came to a stunning realization. There were people in that funnel who were perfectly happy calling themselves fans of the sport, buying merchandise, and consuming F1 content, who might never sit through a full Grand Prix.

"Some people will inevitably not get to that stage," Holmes says. "But it's up to us to make sure that they have the opportunity to interact with the sport in lots of different ways."

Those who did flip from on-demand Netflix soap opera to live F1 racing soon found that the broadcasts were changing before their very eyes. What had been quite old-school racing programs, focused on track position and the analysis of a handful of experts, were soon flooded with new access and data. All the things that Ecclestone had deliberately kept off the air—such as footage from inside the garages and radio communications—were now destined for the screen.

The man in charge of that project was a gregarious Australian acolyte of Rupert Murdoch who'd spent thirty years in the business of making sports look good on TV. His name was David Hill. He liked to joke that he'd saved Kerry Packer's ass when he revolutionized cricket for television in Australia in the 1970s, saved Rupert's ass when he transformed the English Premier League on British television in the early 1990s, and then saved Rupert's ass for a second time when Fox acquired NFL rights in America in 1993. Now he was going to do the same for his old buddy Chase Carey and Formula 1.

Hill had known Carey since the earliest days of Fox Sports in the United States, where they worked together on a couple of other radical reappraisals of the sports-viewing public. One was the bet that American sports fans were much more like Europeans than anyone had imagined. Specifically, they were fractured and heterogeneous in their preferences. The United States wasn't a baseball country or a football country or a basketball country—it was the Nebraska Cornhuskers, the Chicago Bulls, the Seattle Mariners, and dozens of other teams with fervent followings in their immediate vicinities. Hence the birth of the Fox Sports regional networks. Not only were they a new way of catering to fans in specific markets—they were also an end run around ESPN, which Fox Sports didn't think it could take on at a national scale.

The other Hill-led project at Fox Sports that changed his view of the American fan was the launch of an entire channel devoted to another sport everyone told him the United States didn't care about. "When we started the soccer channel," he says, "everyone laughed at us."

Not Carey, though. Remembering their transformative double act, Hill was one of his first calls after the Liberty takeover of F1. Yet again, they were going to repackage what struck the world as a distinctly European product and try to sell it to Americans who seemed to have no clear desire for it.

The first problem Hill identified was that F1 looked like an engineering project, not a TV show. It was uninviting to anyone who'd never seen it, impenetrable to casuals who dared to try it, and unwatchable to him. Which didn't feel quite right, since Hill's initial impression of F1, when he first produced a race in Australia in the mid-1980s, was that this thing stood for absolute excellence, performance, and perfectionism. Why weren't they selling that image instead?

"When Formula 1 arrived, it was like an alien culture," Hill remembers of his first Grand Prix. "I knew that even the guys driving the trucks were the best truck drivers in the world."

Now it was his responsibility to communicate that to an audience that had been numbed by years of processional racing and cars going round and round. That started with access. All of the information that Bernie had guarded with such jealousy needed to be out in the open and onscreen. "There are no inner sanctums," Hill told his staff.

Hill demanded more cameras on the circuits and on the paddock. Believing that sound design elicited more visceral reactions than any picture ever could, he also ordered all the mics turned way up for the first few laps of a race, a little trick he'd developed producing NASCAR for Fox. Only if fans heard the noise of the start could they feel the rumble of twenty Formula 1 cars tearing into the first corner. And most important, Hill wanted live timing and the race order onscreen at all times so that anyone tuning in could understand what in the world was going on.

Bizarrely, that last point wasn't the easiest sell. Hill knew from experience how divisive onscreen ornamentation could be. Soccer fans had hated the introduction of a crazy new gimmick—the game clock—on the early Sky Sports Premier League broadcasts in the 1990s, but not as much as football fans despised a similar innovation for the NFL that Hill called the "Fox box," which informed viewers of the score of the game and how much time remained.

"I received five death threats," Hill says.

Far more popular was Hill's addition of the onscreen first-down line to football games, which helped show viewers exactly what a team needed to do on every play. Showing people where each driver was in the race and the gap to the man in front of him (and maybe their tire choice for the petrolheads) was the Formula 1 equivalent.

Hill even gave the whole show a new soundtrack. For decades, the theme most closely associated with Formula 1 was a guitar lick around three minutes into "The Chain," a 1977 Fleetwood Mac song selected by the BBC. Hill thought that was music for dads. New fans needed the big stirring sounds of summer blockbusters and high-end video games. So he hired a composer named Brian Tyler, who had scored one of the *Fast and Furious* films and a game in the *Assassin's Creed* franchise. The vibe, Hill instructed him, needed to be "Impending doom." The inspirations: a Sean Connery submarine movie called *The Hunt for Red October* and the national anthem of the Soviet Union.

Listen to it for thirty seconds and you'll know that Tyler checked all of those boxes.

Soon, the entire feel of watching a Grand Prix was changing. F1 began to look different and sound different. If the sport was ever going to break out of its reputation as a Sunday afternoon snooze, this radical refresh of its broadcast philosophy would get it most of the way there. Underpinning all of it was a fundamental shift in focus that F1 had never really considered in its first sixty-seven years of existence. And it came down to telling production crews what was really important here.

"Under Bernie, their job had been to follow cars around tracks," Hill says. "But the drivers are the stars, not the cars."

This is what *Drive to Survive* came to understand too. The club of current Grand Prix drivers has only twenty members. Fewer people on this earth get to say they race Formula 1 cars for a living than there are current New York Yankees—and no Yankee is putting his life on the line every time he goes to work. Hill wanted to remind the world that these young men were gladiators. And luckily for F1's business, most of them happened to be pretty good-looking too. So everything about the broadcast was redesigned to put the drivers' billboard smiles onscreen as much as possible. Deep down, people connected with people, not 1,000-horsepower machines and a shade of paint. The relationships they felt with the drivers were what would keep audiences coming back. It was no wonder that inside the Liberty motor home, execs would go nuts anytime they saw a driver walking around the paddock with his helmet on. Their TV show needed faces.

Liberty soon had proof that those instincts had been correct. Ratings started to tick up almost everywhere on the strength of fresher broadcasts and the *Drive to Survive* power boost. But nothing encouraged them more than seeing the numbers in the one market that CVC had failed to crack and that Bernie Ecclestone had decided wasn't even worth the trouble.

FORMULA 1 HAD SPENT YEARS BEGGING FOR TELEVISION COVerage in America. The sport had been through the wilderness of tape delay, cable TV obscurity with the Speed Channel, and four unfruitful seasons under the NBC umbrella that ended with the Liberty takeover. Understandably, NBC had chosen not to renew its agreement with F1 once the sport was acquired by a direct competitor in the broadcast business.

In 2017, the situation was so grim for F1's prospects as a viable television entity that when Liberty struck a deal for the races to be carried on ESPN, it gave away the rights for free. The sport was that desperate for a US foothold. Over the course of the following year, its first season airing on the Worldwide Leader, the sport averaged 554,000 viewers a race. That figure steadily crept up in 2019 and 2020, roughly matching the average that NBC was seeing for its popular coverage of the English Premier League. The difference was that the Premier League played every weekend and occupied at least five time slots. Formula 1 showed up once every two weeks and was often over by people's second cup of coffee.

Drive to Survive, the world's glitziest infomercial, changed all of that. When the third season of the show was released in early 2021, a year into the pandemic and just as F1 kicked off another lap around the world, the show immediately went to the top of the Netflix global charts. ESPN's ratings for Grands Prix surged right behind it. Races that season averaged 949,000 viewers, up 56 percent from the previous year. In 2022, the number was 1.21 million.

It was precisely the validation Liberty needed to pursue its hell-for-leather push into the US market. From 2012 through 2019, the sport's only physical presence in America had been the Austin Grand Prix. Attendance for the race was fine, with around 250,000 fans flocking to Texas for a three-day spectacle that felt like a music festival. But F1 struggled to see much of a future there. "Just a couple of years ago, we were thinking, 'Do we need to stay? Is it worth it to invest?'" Domenicali recalls.

Part of that was F1's own fault. The sport was "too arrogant," Domenicali says. "We were proposing a product and we were just saying, 'This is it.' Coming here, talking about Formula 1 just for three days, and then . . ." He makes a flatline sound.

Liberty had one last push in mind when it picked up the series. As early as 2017, before the Netflix bump and unsure that the project would ever see the light of day, it set the wheels in motion to host a Grand Prix in Miami. The original idea, which picked up where CVC left off, was for a 2.6-mile circuit that took in the palm trees of Biscayne Boulevard and offered soaring views of the bay and of South Beach. But City Hall scuppered that idea when local residents complained about the traffic and the noise and the general inconvenience of handing over their city to spoiled Europeans for a weekend—Miami was all too familiar with that.

Plan B was a little less sexy. F1 partnered with Stephen Ross, the owner of the Miami Dolphins and an active promoter of summer tours by European soccer teams, to design a circuit around his stadium. Though Sean Bratches had initially told Ross that he wasn't interested in the location, two years of administrative roadblocks softened his position. Liberty was so determined to make it happen that there were even reports of it waiving Miami's hosting fee. Ross could offer up the stadium and all of the land around it to organize a Grand Prix with minimal fuss from the not-quite-Miami city of Miami Gardens. Local residents still sued to block the race on the grounds of excessive noise, but failed to make a compelling case that three days of motor racing in their neighborhood would inflict the requisite level of aural distress.

Instead, F1 was free to have its over-the-top US extravaganza, its Monaco-in-Miami. After the pandemic, the years of American indifference, and the revival through *Drive to Survive*, this race in 2021 suddenly felt like Formula 1's American coming-out party. The sport used to be only "for petrolheads," says Laurent Rossi, the CEO of the Alpine team, "a Champions League of engineers . . . People were ready for the show business side to come back."

Miami was tailor-made for the Netflix fan. It still kept all the car stuff—that much would always be there—but everything the series had promised them in the streaming fantasyland would now come true in real life.

"We are in our job interview phase in America," Domenicali said.

Some of the toughest questions in that interview, Formula 1 had never seen coming. Exploding in America when it did—and growing its presence in the world of social media—forced the sport to confront issues it had never fully considered before. In 2020 and 2021, the Black Lives Matter movement swept through professional sports on both sides of the Atlantic. F1 handled it about as smoothly as you would expect from one of the whitest sports in the world.

In July 2020, at F1's first race back from the pandemic hiatus, Lewis Hamilton led a group of thirteen drivers in taking a knee before the start in an antiracism display. But six of his rivals, including Max Verstappen, Charles Leclerc, and Kimi Räikkönen, refused to join the sport's only Black driver and settled for wearing T-shirts that said "End Racism" instead.

Hamilton continued taking a knee before Grands Prix, just as other athletes did from the NBA to the English Premier League. And later that season, after yet another incident of police violence against Black Americans, Hamilton brought the real world into the F1 bubble again by wearing a T-shirt that read "Arrest the Cops Who Killed Breonna Taylor" on the podium of the Tuscan Grand Prix. On the verge of his seventh world championship, Hamilton had finally found his purpose.

"I've been wondering, 'Why me?'" he said. "Why am I the only Black driver that's got through to Formula 1, and not only that but I'm at the front? There's got to be a bigger reason for me being here."

Formula 1 preferred it when he kept that reason away from the circuit. Immediately after the race in Tuscany, and right before the Russian Grand Prix in Sochi, the FIA's race director, Michael Masi, issued a directive banning precisely the kind of shirt Hamilton had worn. Throughout the post-race procedures, the order read, the top three drivers "must remain attired only in their driving suits, 'done up' to the neck, not opened to the waist."

Hamilton shrugged it off.

"They've changed a lot of rules after a lot of things that I've done," he says.

The episode was a rare fumble for F1 under Liberty, which had moved away from the edicts that were commonplace under Max Mosley and Ecclestone and generally encouraged drivers to let their personalities shine. Bernie-World had been team-focused. As the keeper of an uneasy truce between three warring factions—the teams, the FIA, and F1 as the sport's promoter—

Ecclestone didn't have much time for the off-track interests of individual drivers. They were all replaceable anyway. But Liberty had made superheroes out of them, putting their protagonists at the heart of an F1 Cinematic Universe, all in the service of the fans.

"It's an emotional connection," Domenicali says. "Our drivers—there are only twenty, by the way, not hundreds like other sports—are the jewel of our business."

Formula 1 had stumbled on Netflix and learned all about the entertainment aspect of this business. Yes, this was the pinnacle of racing and the cutting edge of automotive engineering. But really, their game was drama and peril and glory—that's what the new audience was hooked on. And as it headed into 2021, Formula 1 was about to deliver a season that even Netflix couldn't have scripted.

15

Abu Dhabi, 2021

IN DECEMBER 2021, TOTO WOLFF LANDED in Abu Dhabi unable to shake the feeling that something was about to go horribly wrong.

Which was strange, because over the previous few weeks everything about this season had been going right for Mercedes. Lewis Hamilton had trailed Red Bull hotshot Max Verstappen by 19 points in the world championship standings with just four races remaining, but he'd closed the gap to nothing with a run of three dominant victories in Brazil, Qatar, and Saudi Arabia. The Mercedes W12 was purring again. And all Lewis needed now to win a record eighth world championship was to finish the last Grand Prix of the season ahead of Max. Mercedes and Hamilton were back in control of their own destinies.

Still, Wolff was nervous. In the back of his mind, he had a nagging sense that having the best car on the grid and the most in-form driver in the sport might not seal the deal. He'd been around motor racing long enough to know that showdowns like this one were rarely decided by which driver crossed the line first. Senna, Prost, Schumacher . . . The history of Formula 1 was littered with final-race shenanigans.

And this had already been a season with more shenanigans than most. At a rain-soaked Spa-Francorchamps in Belgium that summer, the Grand Prix had lasted just over two laps, held entirely under the Safety Car. No actual racing had occurred, yet the baffling decision by FIA race director Michael Masi to send the cars out at all enabled the organization to award points for the race and allowed Verstappen to stretch his advantage over Hamilton.

Then in Saudi Arabia, the FIA again stole the headlines following the third collision of the season between Max and Lewis. In conversations broadcast live on television, Red Bull's sporting director, Jonathan Wheatley, was heard negotiating with Masi over whether Verstappen should give track position to Hamilton or incur a penalty. Red Bull's Christian Horner said the bartering was "like being at the local market." F1 rules were complicated enough to begin with. Now they were being debated and haggled over in real time.

Wolff worried that Abu Dhabi might be heading for controversy too. So just in case, he'd flown a specialist attorney to the Gulf to assist Mercedes should any sort of dispute arise. He hoped it wouldn't come to that. And in order to make sure that everyone was on the same page about giving Formula 1 the good clean fight it deserved, Wolff knew there was one man he needed to speak with.

On the Wednesday before the Abu Dhabi Grand Prix, Toto invited Michael Masi to lunch.

EVER SINCE HE WAS A SIXTEEN-YEAR-OLD VOLUNTEER RUNning coffees at a local racetrack in Western Sydney, Michael Masi had wanted to work in Formula 1. To get there, he spent the better part of two decades grinding his way through the Australian touring car and supercar scenes, half a world away from the F1 heartland.

But in 2018, he received a call from the FIA's Charlie Whiting, Formula 1's immovable race director since 1988 and a fellow Australian. This was a man as influential on-track as Bernie Ecclestone was off the track. He had been the one to nix installing an extra chicane during the Indianapolis-Michelin fiasco of

2005. He was also credited with huge leaps in driver safety, helping to introduce the protective halo around F1 cockpits after a driver went into an ultimately fatal coma following a crash with a crane in 2014. Now Whiting wanted Masi to be his deputy.

That apprenticeship lasted all of nine races. With Masi installed as Whiting's right-hand man, they prepared to tackle the 2019 season side by side. But days before the first Grand Prix of the year in Melbourne, Whiting was late for a morning briefing. It turned out that he'd suffered a pulmonary embolism overnight and died at the age of sixty-six. "I had lost my mentor," Masi said.

Masi received a battlefield promotion. At forty-two, he was the FIA's first new F1 race director in more than thirty years. But if Whiting was seen as part of the sport's furniture, Masi was a complete unknown—and the paddock picked up on his inexperience immediately. In 2020, he'd been widely criticized for starting a qualifying session in Turkey with a crane still on the track. That same year, in Emilia-Romagna, Masi had also been responsible for letting several cars overtake the Safety Car before a group of race marshals had left the circuit.

Drivers also complained that Masi routinely shut them down in conversation. Even when more outspoken veterans, such as Lewis Hamilton or Sebastian Vettel, made suggestions in pre-race meetings, they found him uninterested in what they had to say.

So with a lunch invitation in Abu Dhabi, Wolff hoped to clear the air.

"This is so important, Michael," he told him. "Listen to the drivers. Be open-minded about things. We need to get it *all* right this weekend. I don't want to patronize you. But from my opinion, you've got to have the drivers on board. Just take that feedback."

Masi told Wolff he understood, that he appreciated what was at stake. If he did his job right, there would be no reason for the two men to speak again.

THAT SATURDAY, VERSTAPPEN LANDED THE FIRST BLOW AND put his Red Bull on pole position in qualifying. Then on Sunday, as dusk closed in over the Yas Marina Circuit, Hamilton struck back the moment the Grand Prix started. He gunned his Mercedes off the line and nosed in front going into the first left-hand corner.

Verstappen nearly reclaimed the lead in the hairpin corner at Turn 9, diving through the inside. But Hamilton ran wide off the track and came back on ahead of his rival, having effectively cut the corner. Because he'd been forced off, the stewards determined that no penalty was necessary.

From there, Hamilton was in charge. Midway through the race, his cushion over Verstappen was more than four seconds. And by Lap 50 of 58, the gap had grown to eleven, with five lapped cars sandwiched between them. It looked as if this race was going to end up exactly like the previous three, with Hamilton on the top step of the podium. Even Christian Horner was beginning to accept the inevitable. When the British TV broadcast reached him on the Red Bull pit wall for a live assessment of the race, he had to be realistic.

"The pace of the Mercedes is just too strong," he shrugged. "Max is driving his heart out out there, but we're going to need a miracle in these last ten laps to turn it around."

That miracle came in the shape of a young driver who would soon execute one of the most consequential maneuvers in F1 history. His name wasn't Lewis or Max. It was Nicholas Latifi, and he wasn't very good at racing Formula 1 cars.

The son of a Canadian billionaire, Latifi had been in the Williams seat for nearly two years and scored a total of just 7 points. He wasn't shaping up to add to that total in Abu Dhabi either. Coming out of Turn 14, which had given him problems all weekend, Latifi overcooked the exit and launched his car straight into the barriers. In the race director's office, overlooking the pit lane, Michael Masi leaped into action. He deployed the Safety Car, ordered the marshals to wave their yellow flags, and slowed the race to a crawl while a crane cleared away Latifi's totaled Williams.

Had the crash occurred in the middle of the race, the procedure would have been straightforward. The cleanup would have taken as long as necessary and then the race could have simply resumed. But this was now Lap 53 of 58. Masi knew that there was a real chance that the track wouldn't be clear in time to call in the Safety Car and resume normal racing conditions before the checkered flag. It's not as if he could add on extra laps—F1 cars are only fueled for the exact race distance. So Masi had a decision to make on how this Grand Prix—and the season—would end.

This was exactly what Toto Wolff had been dreading.

INSIDE THE MERCEDES GARAGE, THE TEAM WEIGHED UP HOW to play the final laps. They watched as Red Bull called in Verstappen for fresh tires the moment the Safety Car appeared. If the race resumed, that would give him a clear speed advantage. But Mercedes wasn't convinced there would be enough time for Verstappen to take his shot. Gaming out the climax of the race, Hamilton's engineers felt that it wasn't worth giving up the lead for a pit stop.

The longer the Safety Car stayed out, the closer Hamilton got to his eighth world title. He was out in front, followed by five lapped cars, followed by Verstappen. And as the laps ticked down, Mercedes figured that there were only two conceivable scenarios in play—and neither one could prevent Lewis from winning.

The first possibility was that the race would resume with one lap remaining and five cars between Hamilton and Verstappen. Even Ayrton Senna had never pulled off a lap like that.

The second was that Masi would call on all of the lapped cars—the five between the two leaders, as well as several more farther back—to overtake the Safety Car so that every car on the track would be back in the correct order when the race picked up again. That scenario would put Verstappen right behind Hamilton on newer, quicker tires for the closing exchanges. Yet Mercedes had calculated that the whole process of unlapping the backmarkers, waiting for them to drive all the way around and rejoin at the back of the queue, and then going through a final mandatory lap behind the Safety Car would take too long. By the time Verstappen had his chance to attack, the race would already be over.

Up in the Race Control booth, those same scenarios were running through Michael Masi's mind. On Lap 56, he flashed out an official message in capital letters: "Lapped cars will not be allowed to overtake."

On the pit wall, Horner was beside himself: "They may as well give this fucking championship to Hamilton."

He opened a radio channel to Race Control.

"Why aren't we getting these lapped cars out of the way?" Horner asked Masi.

Masi felt flustered. Supervising the Safety Car, the cleanup of Latifi's wreck, and the order of the cars, all while keeping one eye on the dwindling lap count, he didn't really want to be talking to anyone—especially not this hopped-up Englishman. "Just give me . . . because Christian . . . just give me a second . . . Okay, my main, big one is to get this incident clear."

"You only need one racing lap," Horner pressed.

The trouble was that clearing the Williams off the circuit was taking longer than expected, because the car's brakes had caught fire. At that point, Red Bull's Jonathan Wheatley jumped in with a helpful suggestion: with time running out, Masi only needed to let the five cars between Hamilton and Verstappen unlap themselves—any lapped cars farther back no longer mattered and waiting for them to pass the leaders and circle back around was just a waste of precious seconds. If Masi just did that, then they could resume the race quicker.

"Understood," Masi said, with Wheatley talking over him. "Just give us a second."

"You need to let them go . . . and then we've got a motor race on our hands."

The Grand Prix ticked over to Lap 57 of 58 and Masi made the call that would end his F1 career. He broadcast a message telling the five lapped cars behind Hamilton—and only those five—to overtake the Safety Car. Under the letter of the law, the other lapped cars farther back should have been instructed to do the same. But Masi now felt that would have taken too long.

Even so, Mercedes wasn't panicking. At least not yet. The team knew that Article 48.12 of the FIA rules addressed this exact situation: "Once the last lapped car has passed the leader the Safety Car will return to the pits at the end of the *following* lap." Under these circumstances, that meant the Safety Car would lead Hamilton all the way to the finish line. The championship would be his.

Masi knew all about Article 48.12. But he also knew that his boss, FIA president and former Ferrari team principal Jean Todt, wasn't wild about the idea of Grands Prix finishing under the Safety Car. Though it wasn't strictly against the rules, this was more of an unwritten code.

As they snaked around Lap 57, Masi published one more message: "Safety Car in this lap." There would be racing on Lap 58 of 58.

Now it was Toto Wolff's turn to jump on the radio.

"Michael, this isn't right," he said from the back of the Mercedes garage, his communication broadcast to an audience of tens of millions in real time. "Michael, that is so not right! That is so not right!"

Hamilton was starting to freak out too. As the lapped cars got out of the way, he spotted Red Bull No. 33 in his rearview. "*Fuck,* is he right behind me?" he asked his team. "With new tires?"

Lewis realized what that meant. He understood that no matter how hard he defended, years of F1 regulations had been engineered to encourage overtaking. In any nose-to-tail situation, the advantage always lay with the car coming from behind, especially if it was on newer tires. Hamilton's W12 was a sitting duck.

Verstappen stalked him through the first four corners of Lap 58. Then at Turn 5, he made his move and lunged for the inside. Once he was past Lewis, it was game over.

"This has been manipulated, man," a resigned Hamilton lamented over the radio.

At Red Bull, the team went nuts. Mechanics came spilling out of the garage to greet Verstappen when he took the checkered flag. And Horner held back tears as he shouted into his headset directly into Verstappen's helmet.

"Max Verstappen, you are the world champion! The world champion!"

One garage over, Wolff couldn't accept what was happening. There was no way that what had just occurred on track could be allowed to stand. Mercedes's lawyer was already preparing an appeal. Wolff called Masi again.

"Michael, you need to reinstate the lap before. That's not right."

"Toto," Masi replied. "It's called a motor race, okay?"

"*Sorry?*"

"We went car racing."

AS RED BULL CELEBRATED, WOLFF'S FIRST INSTINCT WAS TO go over Michael Masi's head. If Masi wasn't going to listen to him, he would find someone who would. The FIA reviewed the outcome of races all the time, so Wolff knew that the standings could still be altered if only common sense prevailed.

He picked up the phone and called Jean Todt at home in Paris, where the FIA president was watching Abu Dhabi in the presence of a Canal+ film crew making a documentary about his life.

"*Ah, Toto,*" Todt said to the camera when his iPhone rang.

Wolff explained the situation and told him he planned to lodge an appeal right away. He even had an attorney on hand. The FIA's own rulebook was clear about what should have happened, he said; the Safety Car had come in one lap too soon. But Todt didn't want to hear it.

"The referees must be autonomous," Todt told him. "Have you ever heard [FIFA president Gianni] Infantino say, 'There was a penalty here, but not here?'"

Wolff realized he was getting nowhere. He hung up and marched off to the stewards' office at the Yas Marina Circuit, already pondering Mercedes's legal challenge. He owed it to the team in the garage, the hundreds of employees back in Brackley, and, most of all, to the shell-shocked driver who had just had history ripped away from him. The whole thing felt like such a betrayal that it wasn't hard to imagine Hamilton walking away from F1 for good.

In Paris, Todt's phone rang again. This time it was Horner, who knew the Mercedes appeal was coming. Red Bull would fight if it had to, but as far as he was concerned, there was nothing to relitigate. No one was taking away this championship now.

"I understand there are some protests around," Todt said. "Anyway, what a race."

NOT A DAY GOES BY THAT TOTO WOLFF DOESN'T THINK ABOUT Abu Dhabi.

In the weeks that followed, the FIA launched an investigation into how the closing stages of the most watched race of the season descended into such pandemonium. Its conclusion was "human error." And in January, that human, Michael Masi, was removed from his job and returned to Australia. The FIA, which had reassigned him, said that he hoped "to be closer to his family." Masi refuses to discuss the matter.

By then, Mercedes had already dropped its appeal. Rather than launch into lengthy and expensive litigation that would have damaged the sport's image right as it was crossing into the mainstream, the team decided to focus its efforts on the 2022 season. Even Lewis Hamilton, who disappeared after Abu Dhabi and refused to discuss his F1 future, had been convinced to return.

But for Wolff, those five laps at the end of the season continued to burn. That was the race that ended the Mercedes dynasty.

"When I keep my thoughts running with it, it's so unfair what happened to Lewis and the team that day, that a single individual breaking the rules has basically let that happen," Wolff says, referring to Masi. "Even though he's completely irrelevant: he lives on the other side of the world and nobody is interested in him.

"He was," Wolff adds, "really a total, pathological egomaniac."

And yet even Wolff can acknowledge one immutable fact about his worst day in motor racing. It made for fantastic TV.

In some ways, the Abu Dhabi finale captured everything Liberty wanted Formula 1 to be. The narrative wasn't about engineering or technical specs. This showstopper came with raw human emotion, gripping tension, real controversy, and a deus ex machina that delivered an ending no one saw coming. Cameras from Liberty and Netflix, now with unprecedented access, had captured all of it. In real time, viewers saw and heard the negotiations with Masi, the elation at Red Bull, the outrage at Mercedes, and the quiet disbelief from Hamilton.

The fallout from a crash by Nicholas Latifi in a fight for fifteenth place had given rise to the craziest, most memorable few minutes anyone in F1 could remember.

All of a sudden, any concerns about Mercedes making the sport dull had evaporated. Red Bull had given F1 a brash new champion. And those millions of new fans—the crowd that had tuned in to a random Netflix series when everyone was stuck at home and suddenly found itself hooked on a global motorsport—could be forgiven for wondering: is Formula 1 always like this?

Wolff wished it wasn't. But he understood that his misery was the cost of the highest-rated Grand Prix since the Liberty takeover, with 108.7 million viewers worldwide.

"It's drama and glory, which makes the sport so compelling," Wolff says. "Everyone saw the drama of a worthy eight-time world champion that was robbed of his title . . . I'd rather have it finished the other way around, but clearly that's a mark in history."

In an odd way, Formula 1 had left Bernie-World behind, but this was exactly the kind of headline-spinning uproar that he'd always loved. Even he had never been able to manufacture a controversy like this one. Somehow, a question of how to interpret an obscure clause deep in the rulebook was the incident that launched a thousand pages of coverage. The news cycle stretched out over weeks, filling column inches and hours of airtime, all of it completely for free.

"You know, people spend an awful lot of money publicizing things," Ecclestone says. "And if you can get all this publicity without spending the money, it's a much better thing to do."

Best of all, the audience now got to relive it four months later. Even as it was unfolding, fans online said that they couldn't wait for Abu Dhabi to receive the *Drive to Survive* treatment—even though the new season of the series wouldn't be out until the following spring. The actual race had turned into a trailer for their favorite television show.

Abu Dhabi 2021 had become a cultural touchstone, the only race that most people could name. Speaking at an Ernst & Young awards event in Monaco later, Wolff was quizzed about it for the millionth time. He looked out at the audience of seven hundred and asked how many of them had *not* heard about that race. He counted nine raised hands.

Wolff found that people he met didn't necessarily know about the years of Lewis Hamilton success or what made Mercedes great in the 2010s. But when it came to Abu Dhabi, everyone seemed to have a strong opinion on what went down in those final laps at Yas Marina.

"There were three topics you weren't allowed to talk about at the Christmas table," he says. "Covid, Trump, and Abu Dhabi '21."

16

Lights Out

FORMULA 1 COULDN'T ALWAYS BE that exciting. People had to understand that not every season came down to the final lap of the final race, with everything on the line, while a three-man psychodrama played out over the radio waves. Sometimes an F1 season was kind of boring.

However much the sport had changed, F1's natural equilibrium was to have one dominant team running away from the pack. Someone always seemed to interpret the rules better than the competition, and because the engineering was now so complex, those gaps took years for others to close. In fact, plenty of teams realized that it was often smarter and cheaper to punt on one whole generation of specs and simply wait for the rules to change again.

But to the sport's new owners at Liberty, it didn't seem to matter. F1 was learning to master the trick of being globally accessible to all kinds of audiences, while remaining tantalizingly elusive to experience in person. It always seemed to be just out of reach—too expensive, too fleeting, or happening entirely on the wrong continent. And that only made fans want it more. F1 cultivated that feeling by scheduling live car runs in cities around the world, including Shanghai, Los Angeles, London, and Marseille, like bite-size free samples.

They weren't selling Grand Prix racing, they were lifting the velvet rope just enough for people who'd watched Netflix on their screens to hear the tires screech and engines growl in person. The events were pure fan service, like Comic Con for the *Drive to Survive* crowd, where the heroes signed a few autographs, took a few selfies, and the fans came dressed like mechanics. Just catching a glimpse of Max or Lewis or a snarky team principal was like seeing their favorite television show come to life.

That's why, in early 2023, hundreds of fans lined up outside the Classic Car Club of Manhattan on the coldest day New York had experienced in years. With temperatures in the teens and a biting wind blowing off the Hudson River, they huddled on the edge of the West Side Highway to greet Christian Horner like he was a pop star.

"Normally it's my wife that gets recognized," said Mr. Ginger Spice. "But even the guy at customs was a *Drive to Survive* fan."

Red Bull had come to New York to introduce the RB19, the car that they were all but certain would propel Verstappen to his third consecutive title after following up his triumph in Abu Dhabi with another championship in 2022. By now, the team was no longer counting on human error at Race Control to dominate. Mercedes had responded to the bitter disappointment of Abu Dhabi by pushing all of its chips onto an audacious new design for the W13, only to find that attempting to reinvent the sidepod wasn't one of those championship-winning loopholes. Instead, it backfired in spectacular fashion. The Mercedes bounced up and down for the better part of two seasons while Adrian Newey's latest masterpieces proved unstoppable.

Maybe a little too unstoppable. As Red Bull reeled off race victories that made everyone else look like they were driving shopping carts, the reflex of the F1 ecosystem was to ask the sport's leaders what they were going to do about it.

This time, the answer was nothing.

After Red Bull took each of the first seven races of 2023, Stefano Domenicali came out and assured the sport that there would be no midseason rule tweaks to put the brakes on an unbeatable car running away with the championship. Unlike the Schumacher years, when Bernie Ecclestone and Max Mosley cooked up solutions to keep viewers and sponsors onside and engineered whatever measures they could to encourage overtaking, F1 under Liberty saw nothing wrong with a juggernaut. They were writing a new history for the sport, and if fans were going to look back at legendary campaigns years down the line—the same way current fans can look back in awe at Schumacher—then drivers like Verstappen needed to be allowed to lay down their seasons of unchallenged supremacy.

"If you have a good forward," Domenicali said, reaching for a soccer analogy, "you can't change the dimensions of the goal."

It was easy to understand why Liberty wasn't inclined to change the recipe. Everywhere it went, Formula 1 was breaking its own attendance records. So many circuits were clamoring to host Grands Prix that Liberty planned to squeeze twenty-four races into a single season in 2024 and even considered creating a rotation system to include more venues.

A full quarter of the calendar would now be in the Americas too, moving F1 away from the Sunday morning coffee set and firmly into the weekend mainstream. After seventy years, the sport was closer to breaking through in the US market than it had ever been. Everyone seemed to want a piece of it. Between 2018 and 2022, F1 teams more than doubled the number of US-based sponsors to over a hundred brands, according to analysis by Spomotion. Japanese tire maker Bridgestone was looking for a way back into the sport as well, after more than a decade away. And in 2022, Liberty had to deny rumors that it had declined a $20 billion bid from the Public Investment Fund of Saudi Arabia for the commercial rights to F1. Not only were they unfounded, Liberty said, the price was also a little low.

But nothing underlined the sport's rediscovered legitimacy like the return of the major automotive brands. Only fifteen years earlier, the world's biggest manufacturers couldn't wait to abandon F1. Honda, Toyota, and Renault simply saw no upside to remaining in the sport as factory teams. They felt that supplying the odd engine was a far cheaper way to achieve the same gains in exposure, such as they were. But now major manufacturers were saying they couldn't afford *not* to be in F1.

With new engine regulations on the horizon for 2026, Audi scheduled its return with the dormant Sauber team. Aston Martin agreed to become the Honda works team, while Ford planned to bring back the iconic blue badge as a partner to Red Bull. Demand was so high that there wasn't even room for everyone to hop in. Porsche announced its intention to rejoin the series, but failed to strike a deal with Red Bull. And General Motors also hoped to secure a spot on the grid through a partnership between Cadillac and Andretti Global, but first required F1 to expand the field beyond ten garages. The era of the factory team was coming back.

"A few years ago," Ferrari team principal Frédéric Vasseur says, "we were struggling to get three around a table."

THE MANUFACTURERS' SUDDEN RETURN WAS ALL THE MORE head-spinning considering the global environment Formula 1 was trying to reckon with. By all rights, a gas-guzzling sport that relied on constant global air travel and cozied up to autocrats didn't have an obvious place in the modern world. Sustainability wasn't just a buzzword for F1, it was an existential threat.

The sport responded by saying all the right things. It launched what it called a Net Zero Carbon plan with a target date of 2030. The idea was to develop 100 percent sustainable fuel for use with the next generation of engines, reduce emissions, do away with as many single-use plastics as possible, and seriously reconsider its calendar to minimize air travel. (Despite the perception, only 0.7 percent of the sport's emissions actually come from the racing—most are generated by the enormous needs for freight transport between Grands Prix.) Critics dismissed many of those efforts as "greenwashing," the practice of making it look like a company cares about the environment when it is in fact just reaching for positive coverage. In 2023, they pointed to one stretch of the calendar that went Saudi Arabia–Australia–Azerbaijan–United States. Another was scheduled to string together Brazil, the United States, and Abu Dhabi. And yet despite the cognitive dissonance of racing combustion engines and permanent globe-trotting in the middle of a climate crisis, F1 became the blueprint for nearly every other sport.

Everyone else wanted their own Netflix series. NASCAR, fretting over its primacy as America's No. 1 motorsport, launched a series called *Race for the Championship* on USA Network, a channel known mainly for *Law & Order* reruns. The producers of *Drive to Survive* were hired to work their magic in tennis, golf, track and field, and professional cycling. But without a pandemic to keep people glued to their screens, there had to be more to it than simply pumping out ten episodes of highly polished fly-on-the-wall entertainment. While others were still signing away the rights to the backstage footage, F1 was going places far beyond the garage. Instead of simply documenting what went on behind the scenes, it would create its own universe where racing on a Sunday was only one of the roads into the sport.

"Today, I'm very pleased to see teams, team owners, manufacturers, and drivers all embracing this new way of presenting Formula 1," says Domenicali. "We have the responsibility of content."

More content, for more platforms, for bigger audiences—that was the Liberty formula. There would still be *Drive to Survive* going forward, at least into 2025, but F1 also welcomed another production company to try its hand at taking the sport to the masses. This was a little shop called Apple.

By the US Grand Prix in Austin in the fall of 2022, it was an open secret that Brad Pitt and the Hollywood producer Jerry Bruckheimer intended to get in on the act. Bruckheimer didn't do much to keep a low profile, strolling through the paddock in a denim jacket with the words "Top Gun: Maverick" stitched on the back. And Pitt, as ever, was tailed by a flock of photographers. They were there to meet with F1's principal characters and discuss their vision for a summer blockbuster that was an as yet untitled Brad Pitt/Apple Studios F1 project. More than thirty years after he'd released *Top Gun* and *Days of Thunder* in quick succession, Bruckheimer was going back to the old Fighter Plane–Racing Car double again.

F1 was thrilled to oblige. The sport bent over backward to give Bruckheimer everything he could possibly need. Toto Wolff and Lewis Hamilton signed on as coproducers for no fee. Mercedes built a car for the movie. And Domenicali worked out the kinks so the production could actually film at the British Grand Prix in the summer of 2023, with Brad Pitt sitting in his car on the live starting grid seconds before lights out of the GP2 race. Even *Drive to Survive* had never enjoyed that kind of access.

But as a moviegoing experience, F1 hoped that having its very own blockbuster could finally overcome any lingering misconceptions people had about the sport with pure, uncut entertainment. The noise—a point that David Hill made repeatedly to Bruckheimer—needed to be as loud and as visceral as lying on the tarmac under an F-14. The visuals were so advanced that sitting in the theater would feel like playing a video game.

In the meantime, F1 kept pressing ahead in the world of actual video games. In 2023, Liberty launched its flagship F1 Arcade in London and earmarked £30 million to expand it into a chain of more than thirty locations across the United States, Europe, the Middle East, and Australia. McLaren team principal Zak Brown and driver Lando Norris signed on as early investors. Anyone who ever wanted to sit behind the wheel of something approaching an F1 simulator would be able to. Maybe that could persuade them to tune in on a Sunday. As far as Liberty was concerned, the time to capitalize on F1's popularity was now. And in that global scramble for attention, any angle was worth exploring.

"There are things that make me feel optimistic," Toto Wolff says. "But we've got to fight for our place. You fight for eyeballs and you need to provide entertainment."

THE QUESTION WAS HOW MUCH OF ITSELF F1 WAS PREPARED to give away.

Well, not exactly give away. At the inaugural Miami Grand Prix in 2022, any decent ticket package for the race went for between $5,000 and $10,000. And even that didn't get you anywhere near the real action. That took place inside Lewis Hamilton's garage, where the parade of galactic stars included David Beckham, Michael Jordan, Tom Brady, and First Lady Michelle Obama. Hamilton said the weekend carried "a similar vibe" to the Super Bowl.

Never mind that the actual racing was duller than the card room at a Florida retirement home. The weekend was broadly a hit. But how many times could F1 repeat the trick? By 2023, Miami was the first of three races in the United States—Austin, Miami, and a new event in Las Vegas. Inside F1's offices, they knew that if they didn't milk America now, they never would.

The top-end ticket packages for Miami now started at $10,000. In one hospitality area, a single lobster roll cost $450. Anyone who really wanted to fit in might have sported a limited-edition Formula 1 T-shirt by Chanel with a price tag on the secondary market of around $5,000. (Far cheaper was a squirt of F1's official perfume, with its "fresh, intense, woody fragrance with a metallic twist.") All of it was merely an *amuse-bouche* for the bacchanal coming later that year in Sin City. During a preview event along Las Vegas Boulevard one year out from the Vegas Grand Prix, one team principal gazed out at the Strip and had to shake his head.

"It's like a new sport," he said.

For a certain generation, this was a jarring development. F1 was still trading on its European heritage, but increasingly leaving behind its roots. The new American Grands Prix were crowding out European races. Even Monaco, the quintessential bastion of F1 tradition, began to feel the pressure.

During a tense round of negotiations over the principality's place on the calendar, Stefano Domenicali and Liberty threatened to walk away from the Riviera for good. The Automobile Club of Monaco had heard this sort of thing before from Bernie Ecclestone. The difference this time was that they weren't certain F1's new owners were bluffing. So Prince Albert, Rainier's son, instructed the club to fold. He greenlit an unprecedented set of concessions that all but undid decades of his father's work to preserve Monaco's unique status. The race would no longer be in charge of its own broadcast, nor would it have full control of its own sponsorship. By 2023, David Hill had twisted the prince's arm enough to convince him to allow a helicopter-mounted camera to hover over some of the most exclusive real estate in the world for the purposes of airing a motor race. And even then, the Automobile Club had its doubts that all of this would save its spot in the lineup. Monte Carlo was now just another stop on the Liberty Media circuit, and its new contract ran only until 2025.

"What counts for the Americans is the strength of the offer, not the duration," the Automobile Club's president of fifty-one years told *L'Equipe*. "If a country from the Middle East puts 10 times more money on the table than we do, we're dead . . . Monaco tries to offer a sporting spectacle that perhaps no one else can. If the bosses of the circus don't feel that subtlety, it's obvious that we're done for."

Even Bernie never seriously considered dropping Monaco. Then again, how much did F1 need it anymore when it could just stretch out a sheet of blue plastic near Hard Rock Stadium in Miami Gardens and call it a marina? The very idea would have horrified the old guard that built the sport, guys like Enzo Ferrari, Frank Williams, and Ron Dennis. But Enzo and Frank were both dead. Dennis was in a state of permanent banishment. And Ecclestone was watching every race weekend from his chalet in Gstaad, surrounded by his third wife, Fabiana Flosi; their infant son, Ace; and a collection of British memorabilia that includes ceramic bulldogs and a life-sized replica of a Tower of London guard.

Their moment had passed. The Liberty show was looking for a different type of character.

The days when a used-car salesman or a high school dropout could climb the ranks of Formula 1, buy a team, and alter the course of the sport's history were over. Engineers were no longer tinkerers who monkey-wrenched their way to the top—now they need PhDs just to get in the garage door. Owners weren't hopeless eccentrics, they were billionaires or Hollywood stars. When the Alpine team was looking for investors in mid-2023, it landed on a group that included the actors Ryan Reynolds and Rob McElhenney, who'd successfully deployed the *Drive to Survive* playbook on a struggling soccer team from North Wales called Wrexham. The sport, now covered in stardust, was fully in its Las Vegas era.

"We are a sport, and we obviously have a tremendous legacy in racing, but we're much more than that," says Renee Wilm, the CEO of the Las Vegas Grand Prix. "We are a fan experience. We're about the technology . . . We're about the glamour. We're about the celebrity factor."

That's one reason why the drivers were now selected as much for their crossover appeal as their skill in the cockpit. Liberty was so desperate to see a US driver back in the series—where no American had won a single Grand Prix since Mario Andretti in 1978—that the Williams team reserved a seat for a young Floridian named Logan Sargeant before he'd even qualified for the super license required to race. Sargeant precisely fit the mold F1 was looking for, right down to his fighter-pilot name.

That mold had delivered a field of drivers who were largely indistinguishable from each other, like a parade of former boy-band members with all the right accents. They looked the same, dressed the same, and because so many of them shared the same coach coming up, they drove the same too. The one who stood apart was Lewis Hamilton, just as he always had in an F1 environment with a staggering lack of diversity. He had become the first Black F1 racer in 2007, and more than a decade later, he was still waiting to see the second.

"I thought that [me] being there would break down that barrier and would encourage more people to come through," he says. But when Hamilton found himself looking at team photos on the F1 Instagram account, zooming in, he still saw nearly no one who looked like him. "How is it possible? How has it not changed?"

The irony is that F1 had never before pulled in so many different kinds of fans. The sport's new demographics were younger, more diverse, more American, and included more women than ever before—even if a significant percentage of those viewers had never heard the name Michael Schumacher, according to F1's head of broadcast. The trend was a marketer's dream.

Liberty wasted no time on cashing in. In late 2022, it announced that ESPN had re-upped its deal for the American broadcast rights to F1 for three more years and a significant raise. The company was now paying more than $100 million over the life of the contract, according to a person familiar with the deal—a healthy increase on the original 2017 price of $0.

FORMULA 1 HAS ALWAYS REINVENTED ITSELF, THROUGH teams, through technology, and through creative takes on the rules. But the Liberty era was beginning to feel more like a total reboot.

From their first moments in London, the American executives set out to turn the page on anything that came before 2017. Bernie's reign might as well have been the Cretaceous. A few things still had some commercial value in the interest of selling fans on Formula 1 history—the timeless appeal of Ayrton Senna or black-and-white footage of Juan Manuel Fangio—but if fans thought that F1 began in 2017, then it wasn't the worst thing for them to believe.

Rooting the sport in the now was a bold approach that ran contrary to the choice made by Britain's other great sporting export, the Premier League. English soccer, whose modern era began in 1992, consciously trades on the idea that everything about it has been around since the nineteenth century, from the clubs to the stadiums to the towns they helped put on the map. There isn't necessarily a right answer.

That push and pull between history and zeitgeisty relevance underlies every decision major sports leagues have to make. Do they service the core fans and season-ticket holders who have been around for generations, or do they try to woo a potential global audience many times larger? The difference for F1 is that this dynamic developed almost overnight. The whiplash of a sport once watched almost exclusively by nerdy middle-aged men suddenly discovering that it was cool and young and online was perhaps the most disorienting moment in the series's history. The biggest source of confusion was what exactly all of those new fans were drawn to. Was it the spectacle of high-performance machines tearing around a track or was it the circus that came with it? Was the sport the thing or was it the trappings of the sport?

"There is, of course, the centrality of the sport, the centrality of the fight on the track," Domenicali says. "But we need to be flexible."

The worry in some quarters was that Formula 1 was being a little *too* flexible. Within four years of taking over, Liberty was tweaking the very structure of an F1 weekend. The race still took place on Sunday, but a few times every season, another shorter race took place on Saturday. The so-called Sprint races were designed to give fans everything that regular Grands Prix did not: more overtaking, more chaos, and more time to themselves. For shorter attention spans, they were perfect—particularly since the longer you spent thinking about the Sprint races, the less sense they made: all the risk of an actual race for a fraction of the reward.

That was hardly the most bewildering innovation under Liberty. F1 fans agree that perhaps the biggest swing-and-miss came at the 2023 Miami Grand Prix, where the series debuted NBA-style player intros for the drivers. Moments before they jammed themselves into their cockpits to race at 200 mph, each one was introduced to the Miami public by rapper LL Cool J shouting in their faces and flanked by Miami Dolphins cheerleaders.

"None of the drivers like it," Lando Norris said.

"It's the show," Williams's Alex Albon added. "We're in the show business now."

Every initiative seemed geared to produce more content, one TikTok reel at a time. The races themselves were now being shot specifically to give each fan a moment to lap up. Traditionally, race broadcasts had focused on the drivers jostling for the lead, with the car at the front onscreen for roughly 37 percent of the airtime and scant consideration paid to the backmarkers, according to one F1 insider. Under Liberty, that number was closer to 20 percent for the leaders, leaving more oxygen for the rest of the field. The triumph of *Drive to Survive*—getting viewers to care about the drivers who never won—was now being mirrored in the live broadcasts.

The teams played their part in the Netflix feedback loop too. Being an original *Drive to Survive* star was so valuable that someone like Daniel Ricciardo was worth keeping around despite his not racing a car. When McLaren chose not to renew his contract past 2022, Ricciardo returned to Red Bull as its test driver, but also as its social media jester-in-chief, until he was thrust back into a cockpit at Red Bull's sister team AlphaTauri after its rookie driver proved too incompetent.

Taken individually, each of these choices made sense for the problems F1 was trying to solve in 2017: diversifying its audience and bringing people back to a sport that their dads used to care about. But taken together six years later, they painted a picture of a product that was a little too happy to blur the line between sporting competition and blockbuster entertainment. The drivers felt it too. Max Verstappen, the kid who was born to drive, was barely twenty-five years old and a two-time world champion when he began to think about seeking out a new challenge. Racing wasn't the be-all, end-all he'd grown up chasing.

"Now that I've won a championship it's nice to win another one and another one," he said. "But basically it is the same thing, so it's not something that will keep me here forever."

And if the racing could barely keep its champion engaged, what could it do for someone who'd never experienced the thrill of blowing past a rival or spraying champagne from the top step of the podium?

What Liberty understood was that the racing alone didn't have to. This was the single realization that underpinned every facet of Formula 1's modern transformation from engineering laboratory to twenty-first-century content factory. The desire to race is what created F1, but the racing was no longer the end, it was the means. Some people would never get all the way from watching *Drive to Survive* or liking a driver's Instagram post to sitting down in front of a race for two hours on a Sunday. And Formula 1 was just fine with that.

In that respect, F1 had begun to resemble a *post-sport* sport. It was perfectly possible to call yourself a fanatic of Formula 1—with all the engagement, emotional attachment, and financial commitment of a lifelong supporter—without watching a single race.

It's hard to know if that made F1 a pioneer in an age when old-school professional sports and traditional broadcast were heading for the same iceberg, or if it was merely setting itself up for a crash. Was this the future of sports fandom? Had they cracked the formula, or was this just another loophole waiting to be closed?

Epilogue

Las Vegas, 2023

ON THE CHILLY NOVEMBER NIGHT that Formula 1 realized its $600 million vision for a Grand Prix in the middle of Las Vegas, the scene an hour before the start was everything that Liberty Media hoped it would be.

It was also completely absurd.

The newly constructed paddock, normally a place for scrambling mechanics and dollies loaded with tires, now came with its own red-carpet entrance to welcome A-listers, B-listers, and anyone else who had called in a favor to go behind the velvet rope of the Las Vegas Grand Prix. Paris Hilton, who would never miss a party like this one, made sure the pack of photographers knew she was there in head-to-toe leathers. Shaquille O'Neal, towering over proceedings, didn't have to work so hard to get noticed. Gordon Ramsay, a guest of Red Bull, was one of the few celebrity chefs on hand who wasn't asked to cook. And Justin Bieber, an F1 regular, prepared for his solemn duty of waving the checkered flag. At least a dozen Elvis impersonators of varying girth were on the loose.

All over town, hotels and casinos were offering ludicrously priced ticket packages that promised unlimited caviar, endless champagne, triplex hotel rooms, or all of the above. They came with views of the Strip, fully locked down for the occasion, where the cars would hit top speeds of more than 210 miles per hour as they screamed past those iconic monuments to understated elegance: the Venetian, the Bellagio, and Caesars Palace.

High-rollers ready to shell out $12,000 per pass could secure their perch on the rooftop deck of the Bellagio's Fountain Club, purpose-built to offer spectators a luxury suite experience to rival Madison Square Garden. The menus boasted Jean-Georges Vongerichten's wagyu carpaccio, Mario Carbone's meatballs, and Alain Ducasse's "Parisian chocolates," each hand delivered from across the fountain pool by waiters on a tiny barge.

And yet, in the best Vegas tradition, there was always somewhere more exclusive to be—and for once it wasn't anywhere on the Strip. For one night only, the row of oil-splattered garages and prefab buildings beside the circuit's start/finish straight was indisputably (and improbably) the hottest ticket in the world of sports.

As the 10 p.m. start time approached, the chosen few in the paddock party spilled onto the grid, where mechanics attempted to make final adjustments to the cars while serving as extras in a million celebrity selfies—like trying to do your taxes in the middle of Times Square. In the center of this frenzy, two middle-aged men moved in a pool of calm, wearing triumphant looks and Puma sneakers. F1 CEO Stefano Domenicali and Liberty president Greg Maffei were basking in their moment of glory. They had come to Las Vegas, bet the farm, and somehow not gone bust.

This was F1's largest ever investment in a single race and the fruits of that outlay were all around them. The gleaming new pit building would now be a permanent fixture, the giant F1 video logo on its roof visible from the sky. Liberty had overseen a complete resurfacing of South Las Vegas Boulevard. Drivers' faces beamed down from the side of every tower block. Per Liberty's contract with the city, what happened in Vegas was going to stay in Vegas for up to ten years.

Twelve months earlier, Maffei and Domenicali had stood on a patch of desert and laid down a strip of paint where the finish line would be. Now, in that same spot, a varsity-jacketed, perma-tanned Donny Osmond was about to belt out "The Star-Spangled Banner."

For those back home, F1 had pulled every available lever to make sure that the world could fully appreciate its handiwork on television. That meant bringing back a man who thought he was done with motor racing forever. A week before the event, television sports guru David Hill boarded a short Alaska Airlines flight from Los Angeles to Las Vegas and set about designing the race for TV. He had already brought in an *American Idol* producer to create a Wednesday night opening ceremony that featured a dazzling mix of lasers, drones, and Kylie Minogue.

His next mission was to use the luminescent glow of the Strip to create the most spectacular backdrop the sport had ever seen. But everything that made the Strip a promoter's dream also made it a TV producer's nightmare. With grandstands and tall steel fences installed on either side of the roadway to protect drivers and fans, Hill quickly realized that any shot from the onboard cameras wouldn't look anything like a postcard from Sin City.

"It's like racing in a cage," he grumbled.

No expense was spared to address this issue. Hill needed more aerial shots from more helicopters taking more daring swoops over the course. At any given moment during the race, half a dozen choppers seemed to be maneuvering through the night sky over the racetrack with their cameras trained on 200-mph moving targets. Even Hill, who'd been in this business for more than four decades and had to be roused out of retirement to be here, thought this was pretty special.

One guy wasn't quite so excited about being there. Unfortunately, it was Formula 1's three-time world champion, Max Verstappen. As he lowered himself into his Red Bull RB19, wearing a white, Elvis-inspired jumpsuit, he didn't feel like an F1 legend taking a bold step to promote the sport of his life. Verstappen felt like a clown in the middle of a circus. Truth be told, he didn't really want to be in Vegas at all. The 2023 drivers' title had been secured weeks earlier in Qatar. He'd already beaten his own record for victories in a single year. And by now, twenty-one Grands Prix deep, he thought the season was beginning to drag.

"I don't think, personally, that we need that many races," he said. "Just take all the quality races, I think that should be enough . . . If there will be even more, I don't see myself hanging around for too long."

BY ALL RIGHTS, VERSTAPPEN SHOULD BE THE FACE OF F1's *Drive to Survive* era.

The role he was born to play in this drama was The Natural. Verstappen started driving at the age of four and rose through the go-karting ranks under the strict, often overbearing supervision of his taskmaster father, Jos. "Even if you're winning races and it looks good and everyone is happy, it's never good enough," Max remembered his father telling him. "You always have to work harder."

That kind of pressure would have broken most kids. The deeply stubborn Verstappen persisted anyway. By the time he became the youngest driver in Formula 1 history, it was clear that he possessed something the other children he'd raced against in go-karts did not. Verstappen was a killer who lived to race.

The trouble was that his rise coincided with the arrival of owners who saw Formula 1 less as a knife fight on the track and more as a campaign for eyeballs. They looked at Verstappen, this 1990s baby with a taste for 1980s racing, and saw a budding Gen Z star who could connect with the most sought-after demographics in the business: young men and women with money to burn.

He represented everything that the sport wanted to be today. Max was young, dominant, telegenic, and hilariously unfiltered. Never mind that Verstappen wasn't interested in social media or fashion or scoring an invite to the Met Gala.

Not only did he appeal to the purists for his racecraft and raw speed—he also reached fans that even *Drive to Survive* couldn't pull in. When he wasn't racing his Formula 1 car, Verstappen spent every free moment racing in the virtual world from the simulator in his apartment, streaming live to a quarter of a million followers. His itch to compete was so strong that after winning the 2022

French Grand Prix in Provence, he had only one pressing desire: to catch a short flight back to Monaco, where his idea of a big night in Monte Carlo was locking his apartment door, firing up his three-monitor sim setup, and racing some more against trash-talking teenagers he'd never met.

The irony was that no one in Formula 1 felt more conflicted about the sport's current transformation under Liberty. Verstappen was the ideal star for the American owners right up until he opened his mouth. Throughout the showpiece in Las Vegas, Verstappen took every opportunity to openly hate on F1's shiny new product. He didn't like the track, he didn't like the start time, and he really, really didn't like the circus.

"Ninety-nine percent entertainment," Verstappen said, "one percent racing."

That wasn't what Liberty or Red Bull or the taxpayers of Clark County, Nevada, wanted to hear. But Verstappen wouldn't stop saying it.

In his view, F1 Grands Prix were supposed to start on Sunday afternoons, on sunbaked tracks, with stands full of fanatics who heard the high-pitched whine of V12s in their dreams. A race that wouldn't finish till after midnight local time, with everyone in heavy coats and bobble hats, didn't only feel unusual to Verstappen—it felt downright wrong. The cars wouldn't appear on the Vegas grid until shortly before 1 a.m. in New York, a time when half of Americans, the very supporters Liberty was hoping to court, would already be fast asleep.

F1 said it didn't have a choice. Las Vegas wouldn't let them have the Strip during the day, and besides, the late start helped them make sure fans in Europe and Asia could tune in to this extravaganza. That excuse didn't wash with Verstappen.

"I personally am a bigger fan of just the traditional tracks," he said. "I think we also have to really remember what the core of the sport is, and why people fall in love with the sport. It's not the show. I think it's more the performance of cars and the drivers going to the limit. And that's why, for me, Vegas is cool—but it's not why I grew up wanting to race a car."

Other drivers at least tried to soften their concerns and toe the corporate line. Sure, there were some growing pains, but those were nothing when you considered what the organizers had pulled off to make this happen. When a loose water valve cover on the Strip lost its battle with the tremendous suction generated by Carlos Sainz's Ferrari and tore up the bottom of the car during the first ever practice session on the track, F1 sought to write off the incident as a minor hiccup.

But they were the only ones. The valve cover fiasco meant that Thursday's session had to be abandoned after just eight minutes. And by the time F1 was ready to resume, it was 2:30 a.m. in Vegas and the spectators had all been sent home. The following day, F1 issued a statement not apologizing for what it described as a "disappointing" fan experience, but explaining that it was their only option. By way of compensation, fans who had spent thousands for the privilege of witnessing the inaugural night of action on the Las Vegas Strip Circuit were offered $200 vouchers for the official F1 gift shop, where baseball caps cost $80 and a windbreaker set you back $265.

Unsurprisingly, Verstappen thought this was bullshit.

"If I was a fan," he said, "I would tear the whole place down."

IT WASN'T JUST VEGAS. MAX VERSTAPPEN SPOKE FOR AN EN-tire constituency of fans who felt that their sport had spun off course, becoming less about motor racing and more of a cookie-cutter exercise in corporate branding. The rough edges that they loved—politically incorrect monomaniacs like Enzo Ferrari and Bernie Ecclestone, genuine hatred between drivers, a winking disregard for the rules, and the faint air of true peril—had all been smoothed over.

Change had come so quickly that many of the stops on the schedule now felt interchangeable. That impression was only magnified by the fact that so many races had exactly the same ending. Verstappen's dominance was rewriting the record books. Between the start of 2021 and the end of 2023, the closing shot for two-thirds of all Grands Prix was Max standing atop the podium.

Verstappen could no longer hide that his appetite for this whole thing was waning—except in the offseason, when his appetite was definitely not waning. His penchant for cheeseburgers and soda, which caused him to pack on around ten pounds every winter, was the only thing that Christian Horner could imagine slowing him down.

But the danger for Liberty's business at the end of 2023 went beyond an untouchable team, its dissatisfied world champion, and a whole lot of "back in my day." Even in the afterglow of a successful party on the Strip—which turned into one of the more exciting Verstappen victories of the year and ended with Max crooning "Viva Las Vegas" down his radio—something about F1's dizzying transformation felt undeniably precarious. That's why, on the night of the loose valve cover, Toto Wolff grew unusually defensive when asked whether the episode constituted a black eye for F1.

"How can you even dare to talk badly about an event that sets the new standard?" he said. "You're speaking about a fucking drain cover that's been undone . . . Give credit to the people that have set up this Grand Prix, and that have made the sport much bigger than it ever was."

It's true that more people were watching Formula 1 than at any time in the previous two decades. But that audience brought with it a level of scrutiny that perhaps Liberty Media hadn't seen coming. Each bold move—from new formats to new destinations—now seemed to trigger a referendum on the sport's future, as if one tiny misstep could cause the bubble to burst.

Six years into Liberty's experiment, no one could say for sure what had gone into the magic sauce that put Formula 1 among the great success stories of modern sports. Was F1's rebirth the result of a carefully plotted strategic road map drawn by wily forecasters of the new media landscape? Or was it really Toto's "lucky punch"?

These were questions that no one in Vegas on that November weekend had time to consider, at least not while the engines roared under the neon lights.

The Grand Prix was only fourteen laps old when a new wave of activity buzzed through the paddock. Men in team uniforms appeared from nowhere wheeling huge boxes marked for air and sea freight. They lined them up outside each of the ten garages like roadies dismantling the set at a rock concert. Cases were unlatched. Generators were unplugged. And within six hours, the row of garages would be completely empty.

The hyper-engineered cars in various states of wear were to be broken down, packed in foam, placed into crates, and sent on an eight-thousand-mile journey to start all over again. The next race weekend was only five days and half a world away in Abu Dhabi. The Las Vegas Grand Prix was over before it was actually over, just another stop on the road show.

Acknowledgments

RUNNING A SERIOUS FORMULA 1 team takes a couple thousand people, from the factory to the garage to the pit wall. Writing a serious Formula 1 book often felt like it required just as much as help.

A project of this scope would have been impossible without the immeasurable help, insight, generosity, and humor of Matteo Bonciani. Few people in this business know more about the sport's inner workings—or have a deeper phone book—than Matteo. He shared his expertise in at least four languages (often spoken all at once) and served as a vital guide to the personalities, politics, and many, many acronyms of motor sport.

Inside F1 headquarters, we'd like to thank Stefano Domenicali for a string of always entertaining and informative interviews—and to Liam Parker and David Leslie for making them happen. David Hill, Greg Maffei, Ian Holmes, and Renee Wilm also shared invaluable insight.

F1 is nothing without the teams and we were welcomed up and down the paddock by all of the sport's protagonists. At Mercedes, we are especially grateful to Toto Wolff, Lewis Hamilton, Bradley Lord, and Justin Perras. At Red Bull, Christian Horner, Max Verstappen, Adrian Newey, and David Coulthard all made time for us thanks to help from Paul Smith and Gemma Lusty. At McLaren, Zak Brown and Steve Atkins threw open their team's hospitality. And at Ferrari, Luca di Montezemolo took us through decades of Maranello history in his inimitable style, while John Elkann and Richard Holloway brought us up to speed on the Scuderia's present and future.

We'd also like to thank Adam Parr and Pastor Maldonado for their stories from inside the Williams garage; Stefan Johansson for sharing his experiences with two titans of the sport, Enzo Ferrari and Ron Dennis; Patrick Duffeler who recounted the high-flying Marlboro years in F1, and Nick Clarry and Robin Saunders, who elucidated the business deals that rescued the sport in the early twenty-first century.

Of course, there probably wouldn't be a sport to rescue if it hadn't been for one man: Bernie Ecclestone. He and his wife, Fabiana Flosi, graciously welcomed us to his living room in Gstaad, where he unspooled enough stories over several hours to make this book twice as long. We'd been well prepared to speak with him thanks to Bernie's longtime lieutenant Michael Payne, who had a front-row seat to one of the greatest dealmakers in sports history.

Others who kindly took the time to speak with us include Duncan Aldred, the Automobile Club of Monaco, and the man who remains the last American to win a Formula 1 Grand Prix at the time of writing, Mario Andretti. We are also indebted to the intrepid Rachel Sharp, who tracked down the man who triggered Spygate when he definitely didn't want to be tracked down—and then did it again two weeks later.

For putting this book into the world, we are hugely grateful to the team at Mariner led by Matt Harper, who grasped the vision for this book immediately, trusted us to execute it, and kept us on the racing line from start to finish. Our own supremo and agent Eric Lupfer was indispensable as always in making this project happen at track-record pace, with thanks as well to Kristina Moore and Christy Fletcher at United Talent Agency.

And, of course, there are too many people to thank among our talented colleagues at the *Wall Street Journal*.

Emma Tucker became our new team principal midway through the writing of this book. She offered her support and encouragement—and even contributed to the reporting with a forensic grilling of Toto Wolff in New York.

We owe an immeasurable debt of gratitude to Bruce Orwall for giving us the latitude to pursue coverage of what can generously be described as not his favorite sport, along with Mike

Miller, Cynthia Lin, Jim Chairusmi, and our colleagues on the WSJ Sports and digital platforms teams. We'd be remiss if we didn't also thank Sarah Ball and Chris Knutsen at WSJ. Magazine, as well as our former colleagues there Kristina O'Neill and Magnus Berger, who thought it might be worth doing a cover story on a guy named Lewis Hamilton.

As always, we'd be nowhere without our first readers, Dan Clegg, Ben Cohen, and Sam Walker, and the families who put up with endless talk of monocoques, double diffusers, and quotable lines from Ron Dennis. They are our parents, Lizzie and Anthony, Aline and Jeffrey; Joshua's sister Céline; and Jon's incomparable wife Katie, and his children, Evie and Cooper.

They might not be able to change four tires in 2.3 seconds, but they are more essential than any pit crew.

Selected Bibliography

Allen, James. *2009: A Revolutionary Year*. London: Speed Merchants, 2009.

Bower, Tom. *No Angel: The Secret Life of Bernie Ecclestone*. London: Faber & Faber, 2012.

Brawn, Ross and Parr, Adam. *Total Competition: Lessons in Strategy from Formula One*. London: Simon & Schuster, 2017.

Button, Jenson. *How to Be an F1 Driver*. London: Blink Publishing, 2019.

Couldwell, Clive. *Made in Britain: The British Influence in Formula One*. London: Virgin, 2003.

Donaldson, Gerald. *Grand Prix People: Revelations from Inside the Formula 1 Circus*. Croydon: Motor Racing Publications, 1990.

Folley, Malcolm. *Monaco: Inside F1's Greatest Race*. London: Century, 2017.

Fry, Nick with Gorman, Ed. *Survive. Drive. Win. The Inside Story of Brawn GP and Jenson Button's Incredible F1 Championship Win*. London: Atlantic Books, 2019.

Fürweger, Wolfgang. *Die Red-Bull-Story: Der unglaubliche Erfolg des Dietrich Mateschitz*. Innsbruck: Haymon Verlag, 2017.

Hamilton, Maurice. *Williams*. London: Ebury, 2017.

Hotakainen, Kari. *The Unknown Kimi Räikkönen*. London: Simon & Schuster, 2019.

Irvine, Eddie. *Life in the Fast Lane*. London: Random House, 1999.

Jordan, Eddie. *An Independent Man*. London: Orion, 2008.

Lawrence, Mike. *Colin Chapman Wayward Genius*. London: Breedon Books, 2002.

Mansell, Nigel. *Staying on Track: The Autobiography*. London: Simon & Schuster, 2016.

Matchett, Steve. *The Mechanic's Tale*. London: Orion, 2003.

Mosley, Max. *Formula One and Beyond: The Autobiography*. London: Simon & Schuster, 2016.

Newey, Adrian. *How to Build a Car*. London: Harper Collins, 2017.

Parr, Adam. *The Art of War: Five Years in Formula One*. London: Adam Parr, 2012.

Rendle, Steve. *Red Bull Racing F1 Car: Owners' Workshop Manual*. London: Haynes, 2011.

Rubython, Tom. *The Life of Senna*. London: BusinessF1 Books, 2005.

Skeens, Nick. *The Perfect Car: The Biography of John Barnard*. London: Evro, 2018.

Warr, Peter. *Team Lotus: My View From the Pit Wall*. London: Haynes, 2012.

Watkins, Sid. *Life at the Limit: Triumph and Tragedy in Formula 1*. London: Macmillan, 1997.

Webber, Mark. *Aussie Grit: My Formula One Journey*. London: Macmillan, 2016.

Williams, Richard. *The Death of Ayrton Senna*. London: Viking, 2010.

Williams, Virginia with Cockerill, Pamela. *A Different Kind of Life*. London: Echo, 2017.

Yates, Brock. *Enzo Ferrari: The Man and the Machine*. London: Penguin, 2019.

Yergin, Daniel. *The Prize: The Epic Quest for Oil, Money & Power*. New York: Simon & Schuster, 1991.

Zapelloni, Umberto. *Formula Ferrari*. London: Hodder & Stoughton, 2004.

Index